SP/C240 01013257/10712

D0522015

Making It Personal

Making It Personal

How to Profit from Personalization Without Invading Privacy

Bruce Kasanoff

JOHN WILEY & SONS, LTD

Chichester · New York · Weinheim · Brisbane · Singapore · Toronto

Copyright © 2001 by Bruce Kasanoff

First published in the United States by Perseus Publishing, a subsidiary of Perseus Books L.L.C.

Published 2001 by John Wiley & Sons Ltd,
Baffins Lane, Chichester,
West Sussex PO19 1UD, England

National   01243 779777
International   (+44) 1243 779777
e-mail (for orders and customer service enquiries): cs-books@wiley.co.uk
Visit our Home Page on http://www.wiley.co.uk
                        or http://www.wiley.com

Bruce Kasanoff has asserted his right under the Copyright, Designs and Patents Act, 1988, to be identified as the author of this work.

All Rights Reserved. No part of this publication may be reproduced, stored in a retrieval system, or transmitted, in any form or by any means, electronic, mechanical, photocopying, recording, scanning or otherwise, except under the terms of the Copyright, Designs and Patents Act 1988 or under the terms of a licence issued by the Copyright Licensing Agency, 90 Tottenham Court Road, London, UK W1P 9HE, without the permission in writing of the publisher and the copyright holder.

Other Wiley Editorial Offices

John Wiley & Sons, Inc., 605 Third Avenue,
New York, NY 10158-0012, USA

WILEY-VCH Verlag GmbH, Pappelallee 3,
D-69469 Weinheim, Germany

John Wiley & Sons Australia Ltd, 33 Park Road, Milton,
Queensland 4064, Australia

John Wiley & Sons (Asia) Pte Ltd, 2 Clementi Loop #02-01,
Jin Xing Distripark, Singapore 129809

John Wiley & Sons (Canada) Ltd, 22 Worcester Road,
Rexdale, Ontario M9W 1L1, Canada

UNIVERSITY OF
NORTHUMBRIA AT NEWCASTLE
LIBRARY

ITEM No.	CLASS No.
200484281	658.812 KAS

British Library Cataloguing in Publication Data
A catalogue record for this book is available from the British Library

ISBN 0-470-84396-9

Typeset in 11/17 Garamond by Originator, Gt Yarmouth, Norfolk.
Printed and bound in Great Britain by T.J. International Ltd, Padstow, Cornwall.
This book is printed on acid-free paper responsibly manufactured from sustainable forestry, in which at least two trees are planted for each one used for paper production.

To Kate, a woman of remarkable character and
conviction, who enables me to swing for the fences

Contents

Foreword

Big Brother is almost here. His sister is the telemarketing operator who called you during dinner last night. His nephew runs a sweep-stakes and magazine-subscription service just outside of London.

The same rapid advances in information technology that are pushing businesses into a new paradigm of competition – the one-to-one marketing paradigm – are simultaneously generating more and more opportunities for the abuse of consumer privacy by mass marketers. Making databases of sensitive, individual consumer information available to marketers interested only in next quarter's sales is like providing chain saws to a tribe of slash-and-burn farmers.

We first wrote those words in 1993, in *The One to One Future*. We knew then that personalization and privacy were inextricably linked, but nowhere will this fact become more clear to you than in the pages of the remarkable book you are about to read. Bruce Kasanoff has woven a compelling tapestry of steady, logical arguments, all pointing to the conclusion that protecting your customer's privacy must be the absolute centrepiece of any effort to profit, in the long term at least, from the trend toward increasing levels of personalization. Unfortunately, as he also argues, many businesses are unlikely to recognize this.

No one can deny that personalization is one of the most comprehensive and transforming technological trends in the modern world. You can look around in your own life and see how this trend is changing your life as a consumer. And for the overwhelming majority of people, the change is definitely for the better. Many of us are old enough to remember when:

- To rent a car, you had to stand in line to fill out a contract every time. Even if you rented the same type of car every week from the same company at the same airport, nevertheless each time you had to choose again from the various insurance options, payment plans, and rates.
- To file a claim on your company's medical insurance plan, you had to fill out a new form, diligently entering by hand your name, address, home phone, office ID, and Social Security number in the appropriate spaces. Three claims in a week? That meant three forms to fill out with the same information, over and over.
- To find music similar to the somewhat obscure jazz record you had picked up by accident one day, you either had to be extremely lucky or you had to find a jazz expert lurking among the minimum-wage shop assistants manning the music store.

There is no doubt whatsoever that personalization, as a technology-driven trend, has already improved our lives as consumers, employees, and *persons*. Dramatically. And every day the information technology that makes personalization possible on a broad scale gets better, cheaper, faster.

Of course, the most notable examples of personalization today are on the Web, where computers interact with computers, instantaneously tracking tens of billions of data points, performing complex mathematical analyses before digitally reconfiguring the pixels into a pattern of information designed to appeal to this particular computer user.

Companies treat different customers differently today for one very basic reason: because they *can*. The technologies that enable personalization have only recently become practical, but the fact that personalization can occur now means that it *must* occur. Customers demand it. People demand it.

Our consulting business at Peppers and Rogers Group is based on helping companies come to grips with the business issues and problems that personalization technologies generate. For instance, if a company can now treat Customer *A* differently from Customer *B*, then who at the company should make the decision with respect to each customer? Who should be held accountable for the results? Indeed, if one of the anticipated results is an increase in customer loyalty, then how does the financial impact of this even get quantified?

These issues are complex and difficult enough that Peppers and Rogers Group has built a substantial consulting practice by helping companies deal with them. But without going into the details, one of the most interesting aspects of our consulting business is that the overwhelming impulse of most companies is to view this issue through the wrong end of the telescope. They don't start by asking how personalization and one-to-one marketing can be used to add value for different types of customers, or to improve a customer's life, or to make things simpler, less expensive, or more convenient for the customer.

Instead, most businesses start with the question of how this technology can be employed to extract more money from the customer. How can we sell our customers more stuff, by using the detailed information we now have? In other words, most businesses do in fact regard this new technology in the same way a tribe of slash-and-burners might regard a chain saw.

We hope you don't mark us down as irrepressibly optimistic if we suggest that the only way to begin extracting real value from personalization is to put yourself first in the person's place, rather than in the personalizer's place. Personalization is a tool with which a relationship can be built. Relationships – with customers, with employees, with stakeholders – will create loyalty and mutual value, over a long period of time. This is a fact, although the way in which you choose to try to measure that value is currently an important tactical issue.

In order to create and nourish a relationship with anyone you must, before anything else, earn that person's trust. And the deeper the relationship is, the richer its context, the more trust is required. What does this word "trust" really mean? People trust you if they believe you won't act in ways that undermine their own interests, won't exploit them, won't betray their secrets.

As a customer, you expect the companies you do business with to make a profit from your transaction. Just because they make a profit doesn't mean you wouldn't trust them. But if they make an unseemly profit on you, by taking advantage of you in ways that you have no power to influence, perhaps, or by working behind your back without your knowledge, then your trust would be broken.

Violate your customers' privacy and you violate their trust. It's that simple, right?

No, it's actually not at all simple. Just as different customers have different needs from your business, different people have different levels of sensitivity with respect to protecting their own privacy. In fact, just agreeing on a definition of privacy protection can be problematic. Do you partner with other companies to render services to your customer? Are you accountable for a partner's privacy policies? If you give a customer's personal information to an outside company, is that a violation of the customer's privacy? What if you give it to another division within your own firm? What if you give it to an outside company that you are partnering with to render services to your customers?

And just what is the *damage* of a privacy violation, anyway? If someone steals your car, you can calculate the actual monetary cost of that crime. But what is the monetary damage of getting phone calls during the dinner hour? Or just a few unsolicited emails, for that matter?

Regulatory authorities are taking an increasing interest in these questions. And while this book is not, strictly speaking, a primer on the regulatory issues involved, it's wise to remember that the legal issues are different for different types of information (medical, financial, children's, and so on), different regulatory authorities, and different countries.

So no, it's really not at all simple.

No one can enjoy the benefits of personalization if he or she is not willing to share the personal information necessary to make those benefits possible. And yet, by sharing that information, the person is risking his or her privacy in the bargain. This is the central quandary of the book, a quandary for our age.

You won't find that this quandary is resolved within the pages in front of you. What you will find, however, is a stunning amount of straightforward and clear thinking about it, along with a large dose of practical advice for dealing with it.

What you'll get in this book is insight into how you can put your business on the right path, toward generating your customers' trust. Read through these pages and you'll be given a great education in how you can indeed *profit* – not just by offering more personalization for your customers, but also by protecting their individual privacy.

But first, let's everyone put down our chain saws.

Don Peppers and Martha Rogers
May 2001

Acknowledgements

I n my early days as a partner at Peppers and Rogers Group, I spent vast amounts of time with Bob Dorf, then president of the firm. Bob's endless supply of energy, talent, and supreme self-confidence built the firm and enabled me to get one of the best seats in the house as the world's economic engine began to shift from mass production to personalization. I learned an immense amount from Bob, and still do. Blessed with one of the quickest minds in existence, he is a good friend and a great mentor.

If you can find a dictionary that includes "think big" as a listing, the definition is sure to read "Martha Rogers". Many of us believe as children that we will change the world, but Martha is one of the few who actually has. She accepts no limits on what she is capable of achieving and inspires those with whom she works and the hundreds of thousands who have heard her speak to do the same. Living with one foot in the business world and the other on the academic side, Martha continues to not only drive business change but also to help nurture a new generation of leaders who bring with them a richer view of business relationships.

When I grow up, I'd like to be Don Peppers. He never stops thinking, but only pauses occasionally to think about how best to convey his ideas to others. I've never been more scared – or ultimately more proud – than when I first had to fill Don's shoes delivering an intensive two-day workshop at Hewlett-Packard. Seldom do you get to study another person's capabilities and actions so intently, and I have learned an enormous amount from Don.

Together, Martha, Bob, and Don have made it possible for me to write this book. I can be single-minded, unyielding, and occasionally tactless in the pursuit of new ideas and constructive change, and these three have handled my flaws with a mix of tact, encouragement, and patience.

John Moseley brought great energy and many insights to this edition for the UK. Rob Robertson lived up to his billing as a talented editor turned agent. I never would have found Rob if not for the networking abilities and kind thoughtfulness of Jonathan Ewert. Chris Locke has been a supporter for some time.

I owe Jim Sterne numerous favours. He has been more than generous in sharing his wisdom as a successful writer and speaker. George Day at the Wharton School has been equally generous with his insights, and it has been a privilege for me to have the benefit of George's professionalism and support.

Friends and longtime colleagues at Peppers and Rogers Group have provided encouragement, ideas, feedback, and party invitations – the latter especially appreciated after long days alone in a room. In particular I want to thank Trish Watson, Steve Skinner, Chrisanne McCoy, Elizabeth Stewart, Diane Kroll, Tom Niehaus, Jennifer Makris, Bridgit Lent, Brian Roberts, Kathy Kavicky, Allison McCoy, Elizabeth Rech, Tom Spitale, Corliss Brown, Drake Smith, Marijo Puleo, Lane Michel, Julien Beresford, Terry Kirkpatrick, Weslyeh Mohriak, and Stacey Riordan. Special thanks to supporters elsewhere who have helped me push the boundaries of personalization, including Jeff Nixon, Brian Henley, David Nour, Charlie Ambuhl, Dan Fagan, Jim Coleman, Debbie Smey, Ron Cox, Tom Flynn, Jim Shaughnessy, and Robin Frederick. Thanks also to Marshall Harrison, Rama Ramachandran, Michael Bane, and everyone at Imperium Solutions.

Toria Thompson, an incredibly intuitive person, played an especially significant role in the early stages of this book. She helped cultivate and develop many of the ideas here and has inspired me to be courageous in my thinking and aspirations.

I'm grateful to all the people who consented to be interviewed for this book, all of whom have far too many demands on their time but nonetheless were more than generous with their time.

Michelle Sawyer, our babysitter and household organizer, served as a tactful gatekeeper enabling me to remain close to my kids but still find quiet time to write. Matthew, my youngest, visited often to share with me his latest new words: book, eat, bye-bye, and more recently, work. Jeffrey, who at seven is already focused on writing a book of his own, inspired me with the brilliance of his writing and kept me suitably humble in the knowledge I'm not even the best writer in my own house. Alyssa, who has a beautiful spirit and boundless creative talents, kept me optimistic and always eager to take breaks, especially on snowy weekends. My wife, Kate, is both my toughest critic and most loyal friend. When she likes my work, I know that it will resonate with business leaders and those outside of my network of personalization enthusiasts. I can't begin to comprehend how much she has enriched my life.

To the degree that I've used business jargon or engaged in overly complicated arguments, it's not the fault of my brother Larry, who has constantly urged me to talk in plain language.

Despite the support of all these people, any mistakes of judgement or execution that remain in this book are mine alone.

Finally, thanks to my parents, Pearl and Dick Kasanoff, no longer with us in person but always in spirit, who taught us we could do anything we set our minds to do. You were right.

Bruce Kasanoff
June 2001

Introduction

I magine that your competitor knows half the people in the world: their likes, dislikes, goals, and ambitions. When they have a problem, your competitor is there to fix it. When they spot an opportunity, your competitor is there to help them reap the benefits. If they need more personal space, your nemesis protects them from intrusions. If they get lonely, your adversary helps them to connect with others. This approach isn't reserved just for customers. Your competitor has built a culture that extends the same collaborative approach to employees, suppliers, and partners, as long as each of these respond by helping your competitor thrive.

How would you compete against such an effective, personal approach? By raising salaries? Spending more on advertising? These actions would be futile, because none are tailored to what people need, which is help with whatever challenge or opportunity faces them right now.

This is the promise of personalization, which is what I call it when companies use technology to treat individuals like, well, individuals.

I am *not* talking about the old-fashioned personal service people used to get from the corner shop. That was feeble by comparison, and it only worked when you wanted to buy something. I am talking about personal treatment that is one million times smarter – because it is enabled by computers – and that impacts every role in a person's life: employee, parent, supplier, customer, partner, investor, student, patient, caregiver, and friend.

It is a strange time. Privacy advocates are up in arms, but most people are still more upset over what companies forget about them than what they remember. (It is utterly annoying to have to repeat the same information again and again.) The new personalization solves this frustrating problem, because it boils down to this: companies will remember what individuals like and don't like. When this is done properly, with discretion and respect for individual privacy, it makes life immensely easier for people. As a result, people will be fiercely loyal to whatever company treats them in this manner.

Of course, companies aren't motivated just by altruism. They'll also remember which individuals are lazy employees, problematic customers, sickly patients, and fickle investors. This prospect rightly raises the hackles of privacy advocates, but so far people have been unwilling to spend money to protect their privacy. It is something they value, but seemingly only if it's free. Surveys show that people consider privacy to be a major concern, yet if you offer these same people a modest amount of money, or a small discount, many will volunteer personal information.

On the other hand, few people have even a partial understanding of how much of their lives are now being recorded and stored in corporate and government databases. When I give speeches to business audiences, it is not hard to spook most participants simply by describing the current activities of respectable companies.

This is why personalization is hard. Move too fast, and you will alienate the people you seek to please, because you will trample their privacy and threaten their security. Move too slowly, and your competitors will steal your best employees, partners, suppliers, and customers.

Technology is about to set off a race like we have never seen before to lock in the loyalty of the most affluent, talented, insightful, and motivated people. Think of the Internet gold rush as a false start, a sort of warning sign. It's not enough to simply get your business online. You need to be personal, but not too personal. In their rush to get close to people, some companies will make ugly, huge mistakes that will maim their organizations and damage personal lives.

The road to profits is never easy, and this one is particularly twisting and slippery. But you can't hold back and watch the mistakes of early pioneers. Once a company has used personalization to lock in a person's loyalty, it will be near impossible to disrupt the relationship. More than ever before, the riches belong to enterprises both swift and steady.

Make it personal, and you will make it profitable. Eventually, there will be only two types of companies in the world:

- those who treat individuals in a personal manner and make healthy profits;
- those who treat everyone the same and must scratch and fight to break even.

The race to make it personal requires dramatic changes in the way business leaders think and in the way enterprises are organized and managed. Most companies change reluctantly, usually only when there's danger on the horizon and someone is there to say, "Quick, this way!" So first, I am going to terrify you, and then I will motivate you to discover a better way.

I went to elementary school with a girl named Julie whose father had Huntington's disease, an inherited brain disorder, which results in the loss of both mental capability and physical control. Over a period of 10–20 years the person with HD progressively loses the ability to think, speak, and walk.

The children of Huntington's sufferers have a 50% chance of developing it. Julie lived the first half of her life watching her father slowly decline from being a successful, much-loved physician to a nearly helpless person who died incapable of recognizing his children.

The whole time, Julie knew the chances were one in two that she would develop the disease. Even if she escaped this torture, one of her two siblings probably wouldn't. Sadly, Julie didn't escape. By the age of 30, her symptoms were obvious.

Here's my question: when Julie applied for a job in her 20s, or sought a promotion, did the companies involved have a right to know about her family's medical history? Where's the line between Julie's right to privacy and the company's desire to gather personal information that ultimately will make it more profitable?

There's a pattern that's repeated endlessly in the adoption of new technologies. For reasons that have to do with human nature, most companies don't think hard about the implications of new technology until after they have installed it.

A few years ago, I led a workshop for the most important business clients of a major customer-service software company. The participants were managers of large customer-service centres, the operations that tend to employ thousands of people in dedicated call centres throughout the country. The purpose of the workshop was to discuss the experiences of clients like these who had implemented new call-centre software, but were unhappy with the results. Managers described miserable call-centre employees who are supposed to take a lot of calls in a short amount of time. The new software required them to type in much greater detail the comments and needs of each customer, but most companies had not loosened their standards regarding how long each call should take. I'm not sure if anyone said this or not, but the impression I had after the workshop was that scientists use similar tactics to drive rats crazy in mazes with no solutions. Firms were consistently implementing this software without considering necessary changes in surrounding processes, measurements, and cultures.

The flood of personal information pouring into your company is going to require dramatic changes in the way you do business. Few companies have considered in enough detail the extent or implications of these changes. You can't wait until you implement more technology, because the Web, email, and voicemail already represent a critical mass of technology sufficient to force new policies, processes, and attitudes.

I'm writing about sensitive topics and don't wish to intrude on anyone's privacy. So I've created a number of fictional stories and examples to bring the most troubling and potentially impactful issues to life. *Everything* in italics in this book is fictional, but designed to illustrate real-life challenges. Although the italicized examples are fictional, they are supposed to be plausible and are based to the greatest extent possible on facts. That is, they are *not* science fiction and should be reasonable representations of actual events already happening, or soon to happen.

To help you sort through these difficult issues and understand how to profit from personalization without invading privacy, I'm going to show you how personalization correctly done actually protects – rather than invades – privacy.

We'll explore the common practices of traditional firms and see why business as usual will cause firms to invade privacy, even though they had no intention of doing so. You'll learn more about the business practices of data-collection firms that sound fine to the professionals that hire them, but terrifying to most other people.

In the process of this tour, you may come to share my opinion that new legislation will change drastically the ground rules for doing business globally, forbidding practices that have been accepted for years. From the extremists who want every transaction to be anonymous to corporate lobbyists arguing passionately for self-regulation and free markets, you'll come to understand better the growing debate between personalization and privacy.

You'll hear from the pioneers who are inventing new personalization technologies and come to understand not only how these tools are being used today, but also how they are likely to be used in the future. The key question here is: Whose agenda do these tools serve? Personalization is still too weighted in favour of corporate interests, and that minimizes its positive impact and heightens the threat to individual privacy.

Anything and everything could be personalized, but that doesn't mean it should. We'll look at ways to differentiate meaningful and appropriate personalization from the rest of the myriad of possibilities. Despite the complexity of new technologies, you'll also come to understand that effective personalization depends on simplicity.

You'll see that many of the rules in this new world are counterintuitive, for example, that large companies are now often better able to deliver personal service than small ones. We'll explore the need for boundaries that not only protect individual privacy, but also ensure that new technologies benefit your firm and result in consistently positive – and profitable – outcomes.

In the closing portions of this book, we'll explore some of the most exciting and troubling ground: the reality that dramatic, unintended results will inevitably grow out of the flood of personal data and new technologies across your firm. You can't prevent this from happening, but the more you are aware of the possibilities, the better able you will be to minimize the unfavorable consequences.

Throughout this book, I'll try to stay grounded in reality and provide you with practical guidance regarding what you should be doing now to profit professionally and personally from these trends. But in the final chapter of the book, I'll loosen the reins a bit and share with you four highly positive visions of the benefits that personalization could bring business, society, and the world. Yes, I think personalization will change the world, and in ways that few imagine today.

I wrote this book for people who work in businesses, to help them understand how technology is changing business relationships and what they should do about it. But if you pass along your copy to a friend who doesn't work in business – a doctor, social worker, scientist, student, or delivery person – I hope they'll enjoy it too, because I've worked hard to show how these issues affect our personal as well as professional lives. People often try to separate their jobs from their lives, but when it comes to personal information, the lines are gone. There is no separation. That's the point of this book.

Get Ready to Invent a New Way of Doing Business ...

Using technology to reinvent business relationships requires the ability to think out of the box, to literally invent a different way for your organization to serve individuals or even other companies. To help you nurture this ability, each chapter closes with a thought exercise. There's no single answer to each exercise, since every business and every stakeholder is different. But I hope that by investing energy in these exercises, you'll come to realize just how exciting are the opportunities we face, and how close behind the opportunities follow problems that cannot be safely ignored. Come up with your own answers ... you'll know when you've found the right one, because your employees, customers, partners, and suppliers will tell you.

1

Why Make Business Personal?

ELECTRONIC BOOK REAL-TIME REPORTING NETWORK
DATE: Today
READER TRACKING NUMBER: B43267
BOOK: Making It Personal ISBN 0-7382-0536-2
READER STATUS: Now reading
READER LOCATION: Page 1
ELAPSED TIME ON CURRENT PAGE: 35 seconds
PAGES READ: 5
AVERAGE TIME PER PAGE: 62 seconds
READING SESSIONS: 1
READER INTELLIGENCE LEVEL: High
PRIMARY MOTIVATORS: Career, societal impact, conversational
interest
MOTIVATION (CONFIDENCE LEVEL): Medium

Someday soon, when you read a book like this on your digital book reader, a record of your reading habits will be created in a company's database. You'll be motivated to allow this intrusion on your privacy by some sufficiently attractive offer. Your reading habits will be analysed, categorized, and compared. You'll be much more aware than you are today how your reading skills compare to other people's, and so will the rest of the world. But you may not think much about this process, because everything in your life will be monitored and analysed in this fashion, unless you take forceful steps to prevent it.

Businesses will get *much* more personal in their interactions with individuals. These interactions will span from the intrusive to the supportive, depending on the attitude of each business. This is not a trend or a business fad. It is not the result of a decision made by a CEO. It cannot be reversed, short of a global disaster. It is the inevitable result of the continuing spread of interactive and database technology. The question is: What do you do about it?

Making it personal means treating different people differently. Even in business-to-business settings, such as when British Telecom is buying a service from Barclays Bank, individuals at both companies influence the outcome of the transaction through their behaviour and values. The more you can understand and accommodate each individual's values, the greater your ability to influence the outcome. Technology makes this possible to a much greater degree than ever before.

In this book, I primarily use the word "personalization," which I intend to describe the practice of companies using information about an individual to change the way they treat that person. Whatever you call it, we won't be talking much longer about personalization as though it were a separate discipline. Everything is going to get personal, and the techniques I describe will become a critical, taken-for-granted element of the way we all do business. You will no more think of ignoring the differences between people than you would think of charging a customer for a product you never delivered. When business nears this level, words such as "personalization" will fade from use. So, too, will personalization as a software category or a trade-show focus. It will be everywhere.

With 15 years of engineering experience under his belt, Roger McAllister was in his first week as director of product marketing for a high-technology company. This was his first meeting with the unit's other senior managers.

"OK, let's review what we know about this group of consumers," said Geoffrey Norton, COO, turning towards the database expert, Michael Golding, who rose faster than you would expect a 6' 6" man to move.

Michael clicked on the projector and started his spiel.

"As you know, we've identified 50,000 consumers as a test group to prove that our one-to-one marketing program works."

"Each of these consumers is in a household that earns over £75,000 per year. They're all two-career couples, which makes it more likely they are time starved. At least one adult in each household has graduated from college, which increases the odds that they are Web savvy, and thus will be easier – and less expensive – for us to serve."

"We've also done an analysis of each household's purchases over the past year, and know which ones have purchased one or more new cars during the past two years, taken expensive vacations, or generally spent aggressively on discretionary luxury items."

Roger was impressed. Maybe marketing was more of a science than he had assumed.

"How precise is your analysis?" asked Geoffrey.

Michael shrugged. "Every transaction we have is accurate, but we aren't able to capture all the transactions. So if we know they bought three cars, it really means they bought at least three cars."

Geoffrey seemed to be satisfied with that and gestured for Michael to continue.

"Of course, we also know their addresses and phone numbers."

Spotting an opportunity to ask an obvious question without betraying his lack of knowledge, Roger asked, "Michael, I didn't hear you mention any details regarding our sales to this group over the past year. Do you have information on that?"

Michael looked confused. He glanced at Geoffrey, who seemed interested in the answer.

"These aren't customers," said Michael. "They are prospects, people to whom we want sell."

It was Roger's turn to be confused. "You have all these reams of information about people who have never bought a single product from us? How do we know so much about them?"

"It's not hard," said Michael, "Just expensive and time consuming. We acquire the information from third-party database providers, and then use our own data analysis and segmentation processes to build profiles of this test group. But under expansion economics, with the right response rates and purchase patterns, this could be extremely profitable."

Roger paused, not sure if he was being really stupid, or if he wanted to push so hard during his first week in a new business. But it was such an obvious point, even a hard-nosed engineer couldn't miss it.

"How do these folks feel about the fact that we know so much about them before they've expressed the slightest interest in our firm?"

There was an extended silence while the people around the table considered this point. At last Michael said, "I don't know. No one ever suggested we should ask them. Besides, if we did, it could just spook them, and we'd be defeating our own purposes."

Roger nodded thanks, and forced a smile on his face, but he was thinking: I want my old job back.

This is one current view of personalization, but far from the complete picture. I call this superficial marketing, designed to generate mass sales from the use of personal information. It may work, but it's not the subject of this book.

Too many companies ignore the personal side of one-to-one relationships. Typically, this means they just collect personal information so they can sell more of whatever products happen to be in their marketing plans. It's as if a husband explained, "Sure, I care about my wife's feelings and our relationship, I just don't want to have to listen to her problems or worry about understanding her needs."

Personalization is not just a topic for marketers, but rather the logical result of technology's impact on *all business relationships*, from those with customers to employees, suppliers, and partners.

In 1993, Don Peppers and Martha Rogers wrote a now classic book called *The One to One Future: Building Relationships One Customer at a Time*. It talked about why every business must establish a Learning Relationship with each and every one of its customers. The idea was that if a company could motivate people to teach it how to best serve their needs, then that firm would be able to serve the customer better than any of its competitors. By making loyalty more convenient for a customer than disloyalty, the company gains a loyal customer, perhaps for life. A simple example of this practice is PC banking. To start paying your bills online, you must first enter the names and addresses of each company you wish to pay. If you like, you can also set up automatic monthly payments for bills that are the same amount each month, such as your mortgage or cable TV bill.

Imagine that you sign up with the first bank to offer online bill payment and go through the lengthy set-up process. A few months later, every bank in town starts offering a similar service. How attractive is the idea of switching? Would you do so to save an extra £1 per month? How about £3? The more work you have invested in sharing your personal information with a company, the greater interest you have in making the relationship work. This is the essence of a Learning Relationship, and the driving factor behind personalization, which Peppers and Rogers refer to as one-to-one relationships.

Done properly, one-to-one initiatives create a win–win situation for a firm and the people who have dealings with that firm. In the case of customers, personalized service saves them time and money and gives them access to better and more relevant information. Firms that develop personalization techniques enjoy reduced costs, increased revenues, and stronger loyalty. These companies are also better able to adapt to changing markets. If companies listen to feedback and react to it in a meaningful way, they are able to change every day. This type of change is less wrenching, and more profitable, than waking up one day and realizing sales have declined by 50%.

Traditional product-driven companies such as Black & Decker and Prudential Insurance have a limited number of products to sell. When they get feedback from customers, they typically use this information to steer the customer towards one of their existing products. "Based on your needs, I suggest our variable rate annuity product," an insurance salesman might say. But only a small percentage of customer feedback actually changes the firm's behaviour. Real one-to-one marketing or personalization means that a company takes customer feedback and uses it to actually customize a service or product for that customer. So while product-driven firms stay rooted producing the same products for all their customers, one-to-one firms change every time they interact with a customer. The better they are able to do this, the closer they come to evolving just as fast as their markets are changing. Technology lowers the cost of both gathering customer feedback and using it to customize a service.

In the last few years, many companies have hopped onto the one-to-one marketing bandwagon and started developing personalization techniques for their customers. However, most firms that say they are implementing one-to-one marketing programmes are doing it for their own benefit, not for the customer's. Here is what many companies are doing:

Computer-aided direct marketing

Remember Michael Golding's marketing plan? His firm is using personal information to attract prospects. Using a combination of advertising and targeted promotions, the company will be focused on selling certain products or simply making more money, not on understanding each customer's needs. Direct marketing – what most people think of as junk mail – was merely annoying when computers were 100,000 times dumber than they are today. Now it's becoming unacceptably invasive. For reasons that I'll explain in this book, the days of using personal information for advertising are numbered. It's likely that this practice will be outlawed or highly restricted by new legislation.

One-way dialogues

In a true relationship, either party can initiate a conversation, and both parties have to be ready to interact and willing to change their behaviour based on feedback from the other party.

Selling out relationships

With growing frequency, companies are being vilified in the press because they have decided to sell information about their customers to other companies. In many cases, such sales represent a complete departure from the purpose for which the customers supplied their information. Network Solutions, the domain name database firm that initially had exclusive rights to maintain the Web's "address book," is selling information about all the customers in its database. "On your mark, get set, go!" announces an ad targeted at direct marketers. Available for the first time ever. Approximately 6 million unique customers, sliced and diced for you to target prospects, learn about a specific audience or retain customers ...

Take this information and run with it."[1] This type of action sends a loud and clear message to customers that the company values its relationship with them less than it does their own wallet.

When we strip away these false impressions of one-to-one relationships, what is left?

First, and most importantly, in any relationship between an enterprise and a person the enterprise must acknowledge that the person has a right to control and access his or her personal information. Without this understanding, there will be no trust and no lasting relationship.

The most valuable information about a person's needs and preferences exists in that person's head. Despite the proliferation of third-party data providers and devices that track our movements and actions, humans are complicated, often irrational, creatures. We are fickle, and our lives take unexpected turns. It is vastly more difficult to serve a person without that person's compliance. People will be less likely to cooperate with an enterprise if they feel unsure about the company's true intentions.

Second, one-to-one relationships work when an enterprise seeks every opportunity to provide a meaningful benefit to individuals, such as saving them time or money, or providing them with more relevant information. These are not the traditional benefits that accrue to companies, but rather to stakeholders in that company, whether they are customers, employees, investors, or suppliers. Most businesses need to think about how their actions benefit the individuals as well as the firm's bottom line. Businesses that do this will enjoy dramatic benefits, but only as a result of satisfying individuals.

Third, this type of relationship only works if the individual provides feedback and value back to the enterprise. Companies exist to make a profit. Somehow, the person must be able to compensate the company fairly for its services.

Finally, true one-to-one relationships are constantly evolving. Companies that are motivated to accommodate the qualities that make each person unique will enjoy the benefits of loyalty. People are motivated to collaborate because they are rewarded for doing so. The result is a series of ongoing interactions that benefit both the individual and the company.

These four characteristics can be applied to any business relationship, whether it involves finance, human resources, operations, product development, information technology, marketing, legal, sales, maintenance, quality control, logistics, customer service, or the loading bay. One-to-one relationships will become increasingly important in all these areas. By understanding these relationships, and how to support them, you'll be able to stay close to the people that bring you business success, and you'll be able to avoid – or at least minimize – the privacy issues that will increasingly plague other businesses.

Memory Is Everywhere

(Obtained from the files of an insurance company)

COMPANY CONFIDENTIAL: INSURANCE RISKS

To summarize the results of Project X9, we have built profiles of 5,000 current policyholders that detail:

- *all the food they have purchased through supermarket loyalty programmes, highlighting purchases that are high in fat and excessive purchases of alcohol and/or sweets;*
- *cruises they have taken that have a tendency to attract people with sedentary lifestyles and/or excessive consumption patterns;*
- *motor vehicle records showing speeding tickets and drink-driving offences;*
- *plus eight other negative lifestyle indicators.*

Based on the above, we have validated our ability to use such third-party information to identify high-risk policyholders. The next step is to decide on the appropriate corrective actions. In our next meeting, we must decide whether to:

1. *adopt an holistic approach and show these policyholders how to lead a healthier lifestyle;*
2. *discourage or reject outright policy applications from people who exhibit these behaviours;*
3. *take no action, due to the sensitive nature of this personal information.*

When you think about how personal a business relationship should be, it's important to understand the preponderance of new technologies that will increasingly track our every movement and interaction. Technology has changed the rules of business relationships, and the laws under which we operate, to a much greater extent than most people realize. Each technology is a potential double-edged sword, which can be used to benefit people, or to work against them. Before you use such a sword, it's critical to dull the edge you don't want to use accidentally.

Memory is everywhere. Nothing short of a global disaster will stop the spread of technologies that make it easier to track the daily actions of people and organizations. We take for granted how quickly voicemail, cashpoints, email, the Web, mobile phones, PDAs, GPS devices, alphanumeric pagers, wireless computers, and countless other microchip devices have permeated our daily lives. Every device adds a layer of memory that didn't exist before, because most are linked to a database owned by at least one company.

Every time a person uses a piece of technology to communicate with others, to save time, or money, or to enjoy special treatment such as a shorter line at the airport, that person takes a risk that the information revealed by his activities will be used against him. Technology remembers what we do, and few people understand the extent to which business – and life – will be drastically different in a society that never forgets than it was in one that forgot 90% of what happened.

Techniques that seem innocuous when applied to customers, such as keeping records of what people buy or who they communicate with on the Internet, are perceived as terrifying when applied to employees. Few employees recognize how widespread employee monitoring has become.

The October 2000 Lawful Business Practices Regulations – part of the Regulation of Investigatory Powers (RIP) Act – made it explicit that UK employers can read the emails and other digital communications of employees without their permission.[2]

The act, in part, says that a company can now monitor and record a communication "in the course of its transmission by means of a telecommunication system, which is effected by or with the express or implied consent of the system controller." To make it easier to read, I've paraphrased the following excerpt from the Act; please obtain legal advice to understand how the Act may affect you. But, in the general, monitoring is now legal for the purpose of:

1. monitoring or keeping a record of communications; or to
2. establish the existence of facts; or
3. ascertain compliance with regulatory or self-regulatory practices or procedures which are applicable to the system controller in the carrying on of his business or applicable to another person in the carrying on of his business where that person is supervised by the system controller in respect of those practices or procedures; or
4. ascertain or demonstrate the standards which are achieved or ought to be achieved by persons using the system in the course of their duties; or
5. in the interests of national security; or
6. for the purpose of preventing or detecting crime; or

7. for the purpose of investigating or detecting the unauthorized use of that or any other telecommunication system.

In other words, there are now a wide range of circumstances under which an employer can legally monitor employee communications, just as employers in the United States have been able to do for some time.

The American Management Association conducts surveys that examine how companies monitor their employees. In 2000, they discovered that "nearly three-quarters of major United States firms (73.5%) record and review employee communications and activities on the job, including their phone calls, email, Internet connections, and computer files. The figure has doubled since 1997, when AMA inaugurated its annual survey, and has increased significantly over the past year."[3]

Matt Kramer of Control Data Systems in Minneapolis, a firm that helps companies analyse their email traffic, reminds us that you can never erase your email messages. "I can take my machine out in the parking lot, run over it in a car, throw it in the bottom of a swamp, never to be recovered again, and it still doesn't matter. Because copies of the email I sent to other people are sitting on servers all over the world."[4]

Given that companies collect data about their employees' activities, let's think about the types of patterns for which a firm could search:

- employees who contact competitors or executive recruiters via email or telephone;
- employees who call their home number more than three times a day more than 50% of the work days, which could indicate a lack of focus on work;
- employees who are failing to contribute an acceptable amount of content to the firm's knowledge management systems.

I've kept this list short for now, just to give you the general idea. We'll explore these implications in far greater detail later in the book. But it's important to recognize that until now, companies haven't had access to nearly as much information about individual employees, suppliers, partners, and customers as they will have over the months and years ahead.

Every day, someone invents a better way to identify, differentiate, interact with, and customize the treatment of individuals.[5]

As I type these words on my laptop, software was just introduced that can determine whether the words are being typed by me, or by some other person using my machine. A technique called "keystroke dynamics" monitors the pressure, speed, and rhythm of a person's typing, which apparently is almost as unique as your fingerprints.

Net Nanny is using this technology, which it calls Biopassword©, a service designed to provide an additional level of security for company networks. It works like this: When you log onto a network with a username and password, the system monitors your typing style. The notion is that even if someone steals your username and password, the system will recognize that someone else is using your password, and it will prevent them from logging on.

This technology is just one example of the field of biometrics, in which machines recognize people using biological traits. Among other characteristics, people have unique fingerprints, voices, facial features, and retinal and iris patterns.

Biometrics has been around for years but has yet to enjoy widespread acceptance. One reason is that biometrics technology tends to be overly complex and a bit upsetting to many people. Fingerprinting, for example, is associated with crime, rather than with good customer service or healthy employee relationships.

Net Nanny's approach is unique because it is a pure software solution; there are no body-part scanners or mechanical devices. "Biometric technologies are becoming a more viable, accepted way of authenticating a user, but in order for them to be widely incorporated and accepted by security professionals and end users, they must be unobtrusive and easy to deploy," said Mitch Tarr, vice-president of strategic alliances for Biopassword.[6]

If people and companies are willing to accept biometrics, life will get dramatically better in many ways. You won't need a driver's licence, credit card, or cash card. Machines will verify your identity based on personal characteristics. Instead of swiping your credit card in a supermarket, you'll just look at the amount displayed, and a retinal scanner will confirm your identity.

No one will be able to steal your credit card. Identity theft will get harder. Queues will get shorter.

EyeTicket Corporation is testing its services at airports in Charlotte, North Carolina, and Frankfurt, Germany. EyeTicket™ allows travellers to check themselves in and board aircraft using iris recognition, without credit cards or other ID, and without standing in line. The system can be accessed by any airline, which means that travellers need to just register once, and then can use the system on any participating airline.

EyeTicket is also being tested at sporting and entertainment events. Instead of buying a physical ticket, people can enter a stadium by looking into the EyeTicket monitor.

Many of the transactions I just described used to be anonymous. Seven of us went to see *Lion King* recently. The theatre has no idea who six of us are; it only knows the identity of the person who purchased our tickets via his credit card. If he used cash, he'd be anonymous, too. But if the theatre used EyeTicket or a similar system, there would be a record of the people sitting in each seat. This information could be used to benefit the individuals involved, by tracking their entertainment preferences and recommending new shows they are likely to enjoy.

On the other hand, information such as the identities of each person in each theatre seat needs to be handled with care. It could reveal an extra-marital affair, or be used as evidence in a judicial case. Neither outcome benefits the people involved.

As others have pointed out, most companies aren't tracking individuals as they surf the Web; they are tracking individual computers. Six people live in my household, and we all use the computer in my family room. There are many "cookies" (data used to track the user) on the hard drive of that machine, placed there by the hundreds of websites we visit. But in most cases, the cookies do not identify a person; they identify the personal computer in the family room. A technology such as Biopassword could someday erase this confidentiality, enabling Web sites to identify each of us via our typing or browsing habits.

Iain Drummond is CEO of Imagis Technologies, a company that offers another type of biometric technology, which is facial recognition. The company's services were initially developed to help law enforcement and security organizations. The Royal Canadian Mounted Police, or RCMP, approached Imagis to help solve a problem. Drummond said, "Criminals, not surprisingly, were often uncooperative when they were being booked. They refused to give their names, or provided false names. RCMP asked us to develop a system that would search a suspect's face against a database of mug shots, which we did."

For over 10 years, Birmingham has been using closed circuit television cameras to monitor strategically important locations in the City Centre. Recently, the city integrated a facial recognition technology known as FaceIt® into its CCTV system. According to Visionics, the company that provides this technology, the system automatically scans the faces of people passing in front of a camera and searches against a selection of photographs of criminals held in a database at the police control room. If there is no match, the facial images are discarded.

A Visionics press release says that FaceIt technology is also at work, "in the London Borough of Newham, where, in association with the Metro-politan Police Service, 300 cameras are tied into the CCTV control room.

After two years in operation, FaceIt was credited with a 34% reduction in crime, prompting a visit by Prime Minister Tony Blair and Home Secretary Jack Straw."

In biometrics, you hear a lot of talk about compliant versus non-compliant subjects. Criminals are non-compliant subjects; they don't want to be identified, will not look squarely at a camera, and will do everything possible to deceive. Compliant subjects are generally people whom you are protecting or treating in a special manner. Drummond describes a bank that recently approached Imagis to help develop a solution to help prevent cash-dispenser fraud.

"Basically," said Drummond, "you stand behind me, see my personal identification number, steal my card, and clean out my account. Using our facial recognition technology and the cameras already in cash dispensers, the bank could confirm that the person who is putting card into the machine actually is the right person. This doesn't add an additional level of surveillance; the bank already knows you're at the cash dispenser, because you are putting your card in. This is a situation in which you are compliant, because you want your money, and you don't want anyone else to get it.

Compliant people will look right at the camera and will have another picture taken when they grow a beard or get glasses. The single greatest factor in determining compliance is whether the person being identified is comfortable with the purpose, method, and end result of being identified. Think about the Imagis technology now being used in casinos. Professional gamblers don't want to be identified, but high rollers do. If the end result of being identified is that you get free drinks and dinner, and a much better hotel room, you will likely be thrilled to have your photo taken.

Compliance is going to be one of those words that you knew but seldom used, until suddenly everyone uses it. Without the compliance of individuals, a firm can barely use the wealth of technologies that enable stronger and deeper relationships between it and the people critical to its ongoing success. The paradox of personalization is that people increasingly expect personalized attention, but if you get too personal, they get scared and want to have nothing more to do with you. It is difficult to know where to draw the line between personalization and privacy, and this is made even more challenging by the fact that every person draws his or her line differently.

Here's the key question: How do you give people enough confidence to be compliant with technologies that recognize and record their typing patterns, entrances into buildings, outgoing emails and incoming telephone calls, and a bevy of other activities?

Companies must be able to explain to individuals that a certain tracking technology is in their interest because it will save them time, or money, or both. They also need to describe the benefits the firm enjoys and to set boundaries around the use of such technologies. Here are some of the ways a company might justify the use of such technology:

- to make sure every employee gets promoted just as soon as he or she demonstrates a certain level of achievement, we need to do a better job of tracking your performance and milestones;
- to give your children every possible advantage, you should consider giving them access to our personalized educational services.

No matter what explanation you provide to motivate compliance, companies must make a convincing case why it is in their interest to use this technology solely in ways that benefit both the person and the company.

People will only feel comfortable with new information technologies when they trust that companies are committed to keeping their information private. People must understand how both they and the companies involved will benefit through the use of new technology. This is an utterly different standard than existed in the past, because people were not being asked to divulge so much information on such an unrestricted basis.

Making It Personal Means Creating a Non-zero Relationship

A wealthy entrepreneur once explained to me, "In every transaction, someone loses out, and I make sure it's the other guy." This entrepreneur is notable more for his frankness than his business approach. Many successful business executives play to win, and their interests clearly come before all others, no matter how polite their conversation.

Too often, business is "played" as though it is a zero-sum game. In such a game, one side wins and the other loses. Ties aren't allowed. We read about strikes to see who won, management or labour? Suppliers negotiate with their biggest customers, and their executives often come back to report that "we got beaten up" in the negotiations. Investors argue with management that unless they lay off 10% of their employees, the company can't succeed. In theory, companies should only compete with their competitors, but that's rarely the case. Most companies struggle against the very individuals that enable them to function.

You can't approach business as a zero-sum game when you are collecting and storing personal information about your most valuable stakeholders. You'll say the right things, but you'll do the wrong ones.

Business becomes a zero-sum game when companies are constrained by limited resources or when they are trading in relatively undifferentiated products. If all you do is sell a product that people can buy elsewhere, you are playing a zero-sum game. The same is true if all you offer employees is a basic job they can find elsewhere. If you are a distributor and you don't add much value, you're also playing a zero-sum game. You're trying to take money out of the pockets of all the other parties in a transaction, and they're all looking at you and wondering what they are getting in return.

By making business relationships personal, a company adds another dimension of value to an existing relationship:

- Customers don't merely get a product or a service. They get convenience that other firms can't match, because the merchant has learned to accommodate the customer's unique needs.
- Employees don't just get a salary and benefits. They get a situation tailored to their skills and a relationship that fits well with the rest of their life.
- Partners don't just get to distribute a firm's products. They get tools and treatment that maximizes their income and cultivates their loyalty.
- Suppliers don't just get a customer. They get a partner who helps them lower costs and risks and also helps them improve their offerings.

Most companies that view business as a zero-sum game will have difficulty aligning the firm's needs with those of their valuable stakeholders. The pot seems limited, because it is. In this situation, talk about "customers come first" or "employees make it all possible" is just lip service. The company and its employees can't both win; there's not enough to go around.

Even in the best situations, companies bring together stakeholders whose interests may appear to be at odds. For example, investors want the highest possible return on their investment, which means that the companies in which they invest must keep costs low. Employees and suppliers are often viewed as part of the "costs" of doing business. Are their interests different than the investors' interests? Unfortunately, the answer is often yes, but not always. Some investors are sophisticated enough to understand that a company can't prosper without motivated, loyal employees and reliable suppliers, so they accept the "reasonable" costs of these parties.

A non-zero relationship is created when both parties seek a win–win outcome and when their situation permits this type of result. This can happen when a company creates an additional source of value, one that exists in addition to the core transactions that take place between a firm and its stakeholders. Take a basic job with an average salary, and add the flexibility that allows a person to accommodate the unique needs of his family, and you create new value without incurring incremental cost. Listen to distributors and use their feedback to change the way you keep them updated about promotions and new products, and you can increase their profits without sacrificing yours.

Here are the characteristics of a non-zero relationship:

- neither party is in sole control;
- both parties define success as a positive outcome for each side;
- important feedback changes behaviour in an immediate and meaningful way;
- both parties are capable of continuing the relationship where the last interaction left off;
- the longer the relationship exists, the more important it becomes to both parties.

Implicit in this process is a greater degree of accountability than previously existed in most business relationships. By becoming more attuned to feedback, and by linking interactions together, a company will become more aware of the behaviour of its stakeholders. The reverse is true for stakeholders. In some cases, one or both of the parties will recognize that the relationship does not have significant enough value to be continued. Here, too, personalization can have a positive impact. Instead of simply terminating an employee who falls short of expectations, a company could better tailor its suggestions regarding the right situation for that person.

Less Is More

Personalization is a more efficient and effective way to do business. It's one of the first ways we've invented to generate higher revenues by doing *less* for a customer, or to use a similar approach to make an employee, supplier, or partner more efficient and effective.

Here are some ways to make business personal, and win by doing less:

Give less

Instead of receiving 134,263 free citations on a search engine, some people will pay a modest fee for a single citation that is exactly what they need. Companies waste tremendous amounts of money by managing information poorly. People have been reluctant to pay for information online, but only because even customized information online still tends to be a few steps away from the level of relevance most executives need. If I could tell you the 17 steps you need to take to cut 20% of the cost out of your most important project – and if I had the credentials to prove you could trust this information – odds are you would pay a reasonable fee for the information. Even better: I give you the code that you can simply insert into existing software. Still better: My system monitors yours remotely and alerts you when it can save you time, money, or both.

Forget less

Instead of forcing employees to fill out the same forms repeatedly, firms can remember their personal information and only require an employee to act when information changes. For example, if an employee has to go to the dentist six times to complete necessary work, the employee shouldn't be required to fill out the same form six times, but instead should be able to submit one form that states six visits will be required. This principle doesn't just apply to forms, it works with test and measurement equipment that engineers use regularly, interactions that employees have often, and documents they need to find. Virtually every employee, supplier, customer, or partner has a laundry list of maddening examples of how large companies make life inefficient and frustrating. How many times have you been asked for your telephone number/employee number/ name two or more times in a row on a single phone call to a support centre?

Promote less

Many firms deluge customers and partners with ads, faxes, emails, and phone calls, often for products that are irrelevant to them. This type of barrage trains people to ignore the company's communications and is generally the result of either ignorance of the other party's needs or an obsession with an internal goal that's viewed as a greater priority than valuing the other party's time and needs. You see this all the time at the end of a sales quarter, when salespeople are pressured to bring in new business, whatever it takes.

Duplicate less

One executive, based in the northern part of the country, relates what happened when he had a rare business trip to his firm's London offices. He discovered that his colleagues there had spent six months solving a problem that his team had solved years earlier. Personalization is about establishing neighbourhoods of people with similar needs and interests and preventing this sort of outrageous waste of resources. Neighbourhoods don't have to be based on physical proximity; they can be based on interests, needs, problems, or opportunities. In this situation, the marginal cost of reusing the original solution is close to zero, but the benefits to the second team are dramatic.

Make less

Companies spend time making products that no one will buy, because the company doesn't know enough about its customers, its partners, or both. Many large manufacturers are ignorant about their end-user customers – they literally can't identify them – and thus it's too easy to miss shifts in customer needs or in the products distributors want to carry.

Businesses need to remember the details that make each person unique. They need to understand better the touch points through which individuals interact with them and look for ways to reduce the costs and maximize the positive results of each interaction. Most of today's business practices have their origins in a time when we did not communicate through computers and when we had a fraction of our current ability to remember, analyse, and leverage information. The goal of companies shouldn't be to collect more information, or to push more services at individuals. Instead, the right goals are to build more trust and more relevance into every business relationship.

Making Things Personal Across an Enterprise

It was a perfect spring afternoon in the suburbs of London, and Fred Jones and his son were washing Fred's new sports car for the second time that weekend. They were working on the hubcaps when Fred's wife brought out the cordless phone.

"It's Jack at the plant again," she said with evident concern.

Fred grabbed the phone, his stomach already tightening. Jack wouldn't be calling unless his solution for the main assembly line had failed.

"Jack, please tell me it worked."

"Sorry, Fred. We tried all the adjustments you suggested. The tolerances are worse, not better. We need another idea, and fast. Every hour we go at this pace costs us £30,000."

Fred walked down the driveway and into the middle of the quiet suburban street. He needed help. He just didn't have enough experience using plastics in this area, and none of the engineers at the plant knew any more than he did.

"Fred, are you still there? My guys are waiting for an answer."

"Jack, go back to the previous method, and give me 60 minutes to check out an idea."

Fred raced into the house and logged onto the company's network via his laptop. He'd been so absorbed in his work the past few months that he hadn't paid any attention to 1ForAll, the new knowledge-sharing tool that connected experts across the 40,000 person global company. He remembered a colleague in Manchester bragging about solving a problem in 35 minutes using the system.

It took Fred about 20 minutes to register for the system, providing details about his current projects and his unique areas of expertise. Despite his urgency, there was no way to skip this step. Anyone seeking help had to first make him or herself available to help others.

At last, he was in. Glancing at the clock every 20 seconds, he gave the details of his problem and selected several keywords from pull-down menus to help the system decide which employees were most likely to have relevant expertise. Classifying the problem as urgent, he submitted his request and instructed the system to dial his mobile-phone number as soon as anyone responded via email. There was nothing to do but wait, so he wandered back out to the driveway and helped his son dry the car and put away the washcloths.

The phone rang. It wasn't the automated voice for which he was hoping. It was even better. The caller was a German engineer working on special assignment for the company in Australia. "Fred," he said, "I have very good news for you. We struggled with the very same problem for eight bloody weeks. But you can solve it in two hours. Grab a pen, will you?"

The next day, Fred was praised as a hero. Then, he shut his office door for an hour and logged back into 1ForAll, giving the system pointers to his most valuable documents, allowing the system to sort through and copy many of his email messages, and signing up as an expert on several topics. From now on, he would automatically be alerted when someone else had come up with an idea that could support his work, or when a colleague needed help in solving a problem that he had seen before.

Fred was so happy about this achievement, it never occurred to him that the price of this new tool was giving up a piece of his privacy.

Companies tend to be organized by functional area, geography, or business units, but life seldom fits well into such neat little boxes.

Technology makes it possible to create connections around any problem or opportunity, and these connections can form and dissolve as needs change. Despite the hype of recent years, we are only just now acquiring the technology to establish reliable connections that link individuals who have one or more needs, habits, experiences, opinions, talents, or situations in common. These people may not know each other or even interact directly. But, in theory, personalization technology can connect them to solve problems or exploit opportunities, by highlighting the previously unknown connections between them.

For example, a technology known as collaborative filtering spots similarities – or differences – between people, products or any other data set, and this information can then be shared with individuals to make their lives easier.

The benefits of personalization will work in every corner of the enterprise. But personalization is a new topic to many executives, especially those outside marketing, sales, and customer service. This isolation has prevented most firms from delivering meaningful personalization to customers and from enjoying the efficiencies and increased effectiveness of personalizing functions within a firm.

Until personalization is an enterprise strategy, no single department or group will be able to deliver meaningful personalization to its stakeholders. If you think that personalization does not apply to your business unit, or your job, then you are wrong. To help you understand why personalization is so important to every aspect of a company, here are some examples of personalization that have nothing to do with marketing.

Employee Relationships

Over the past few years in the United Kingdom, and in many economies around the world, companies have had to compete to attract and retain talented workers. In Silicon Valley in the United States, some companies resorted to recruiting talent in the parking lots of their competitors. Instead of competing on the basis of compensation, it makes more sense to compete by offering prospective employees the flexibility to accommodate their personal needs.

Thanks to personalization, many people may get the dream job they always wanted. Bruce Tulgan, author of *Winning the Talent Wars*, argues that companies should create as many career paths as they have people. "When a valuable person goes to the trouble to customize his work situation, negotiating special arrangements with the organization, his manager, and his coworkers, his stake in the position grows tremendously,"[7] Tulgan wrote. He said that, in today's world, firms should hire for talent, not conformity. It's usually better to have a great person working odd hours from a remote location than an average person who is always in the office.

Tulgan tells a story about delivering a career-success seminar for summer students at Arthur Andersen, where one of the hundred plus participants raised his hand and asked how he could: do work that is really interesting and meaningful, work on his own time, work when he is feeling inspired, work with really smart people whom he likes a lot, in a great company, in a location that is a fun place to work, where he is learning a lot, and making a lot of money really fast.

Tulgan, an expert on new management practices and head of a research and consulting firm, couldn't help but laugh. But then he looked around the room. All the heads were nodding. This is what we all want.

At many companies, uniformity is more important than results. Few would admit this, but it's the inescapable conclusion you reach by examining life in big companies. While some types of business units require uniform or constant coverage – a retail store, for example – most businesses don't need people to work certain hours in certain locations. It's in both the interests of a company and its employees to pay for results, not appearances – if the employee has sufficient initiative and discipline to thrive in such an environment. I'd argue that this is true of far more employees than currently enjoy such flexibility.

Beyond flexible jobs, personalization can enable employees to be far more efficient and effective. A few years ago, I conducted a series of seminars with senior and middle-level managers from Great Plains Software, a company recently acquired by Microsoft that makes accounting software for middle-market companies. Typically ahead of their time, Great Plains executives formed four committees to accelerate the development of one-to-one relationships across their enterprise. One committee focused on partner (that is distributor) relationships, another on customer relationships, the third on employee relationships, and the fourth coordinated these initiatives across the firm.

At the time, Great Plains was my only client interested in building one-to-one relationships with employees, so I asked why this was so. To illustrate, they went around the room giving examples of current practices that wasted resources, aggravated employees, and resulted in frustration on the part of other stakeholders. One manager related that every time an employee called the firm's help desk with a computer problem, the response would be, "What's the port number on your computer?" The manager fumed, "How should I know the port number, and besides, you guys installed it. Why didn't you remember it?"

Managers complained about having to fill out the same information every time they filed an insurance claim. They were upset that there wasn't a better way to learn from each other's successes and failures. We ended one discussion debating whether everyone in the company should get a T-shirt that said: "Never make anyone tell us the same thing twice." In fact, this would be an ideal mantra for any personalization initiative. Mindless repetition is great for learning multiplication tables, but it is a lousy way to run a business.

In fairness, Great Plains Software was – and still is one of the most relationship – oriented, technologically savvy companies in existence. If it found so many opportunities to make employee relationships more personal, you can, too.

You might start by recognizing – and accommodating – differences in the way people communicate and learn. When I worked at Ogilvy & Mather, most executives participated in a half-day brain-mapping seminar that showed us whether we were left-brained or right-brained. The result was a colourful chart that many people posted on their office door or wall. Once you'd taken the seminar, you could glance at a person's chart and know she was a creative artist, not a quantitative manager, so if you wanted to get your point across, you'd better drop the attitude and talk more about emotions, pictures, and people. The most interesting aspect of this seminar was that it wasn't obvious in advance which people fit into which classifications. Some of the most rigid thinkers came from the creative side; some of the most emotional, out-of-the-box thinkers were account managers.

Partner Relationships

As companies accommodate the differences that make each of these stakeholders unique, relationships with partners will change dramatically. What makes distributors different from each other? Here's a sampling. They:

- carry different products;
- charge differing prices, based on changing circumstances;
- maintain varying inventory levels;
- have widely variable abilities to provide customer support;
- have different geographic presences;
- have different credit ratings;
- have their own goals, which bring them in and out of sync with the manufacturer's goals;
- have different abilities to tailor their products and services.

Large manufacturers generally have capabilities their distributors and suppliers lack. One is that they can afford to create tracking systems that no individual stakeholder could afford to develop. Computer-tracking systems are particularly important in enabling the kind of personal treatment that fosters lasting competitive advantage.

Trane makes industrial air-conditioning equipment. Trane typically sells its products to distributors who sell to retailers and repair shops. Personalization does exist, but only in manual, paper-based processes. Trane knows what it sells to each distributor at what predetermined discount and terms, but the company used to have no idea what the distributors were doing further down the channel. Trane decided that it wanted to track order status, accounts payables and receivables, service issues, and support needs.

Using software from Click Commerce, Trane set up a private e-marketplace that automated the entire distribution channel. All the channel partners could then automate the flow of information and colla-borate to get Trane's products out to market faster. Think of Trane at the centre of a hub, with distributors around it, retailers and repair shops around each distributor, and customers around each of these.

In the past, if Acme Distribution sold Trane products to Harry's Repair Shop, Trane didn't know. It wasn't that Trane didn't know about the sale; Trane didn't know Harry existed. Now, Trane knows that the sale took place, how much of a discount Acme gave Harry, when the product was shipped, if there have been any problems, and what level of service Ed has been receiving.

There's a real trade-off here that only works when all the stakeholders have a win–win perspective. Trane gains tremendously valuable informa-tion that should enable it to spot both problems and opportunities in the marketplace. If Trane wants distributors to participate in this system, it must convince them that the system will save them time and money and do no harm to their existing business. Distributors will need to be assured that certain information such as mark-up, accounts, and territories will remain private and confidential. These hurdles are enormous, so a bond of trust must be established. The Click Commerce system that Trane uses enables each distributor to personalize the system. Each distributor sets up the system for their customers.

Distributors are wisely afraid of being cut out once a manufacturer learns about their customers. But that's most likely to happen only if the distributor fails to add value. As long as there is logic for having middlemen, effective ones will be more highly valued thanks to such a system, not less.

Supplier Relationships

In 2000, I had the dubious honour of having one of the first dot.com companies to shut its doors for lack of funding. I had founded

Accelerating1to1 to help companies better measure the effectiveness of their customer-relationship management, or CRM, programmes. We were targeting companies that were just starting to implement one-to-one marketing programmes. One of the things we discovered was that companies were far less advanced in this area than most people assumed. The large companies we approached didn't have difficulty measuring the progress of their CRM efforts; most considered their efforts to be extremely early, and not especially sophisticated. "We know we're at square one" was a typical reply.

But just before we closed the doors to Accelerating1to1, we discovered that many companies were interested in getting better treatment from their suppliers. If the company was a big and important enough customer, it needed to teach its suppliers about customer relationship management.

Here's an example of what I am talking about: Virgin has thousands of employees who use credit cards, work on computers, depend on outsourced help desks, order office supplies, make long-distance calls, and hire consultants. It's in Virgin's interest to request customized service for all its employees. Inspired by our mantra of "never make a customer tell us the same thing twice," Virgin could insist that all major suppliers remember the preferences of each Virgin employee. Suppliers would need to remember the items and services that employees use regularly and to maximize their ability to save employees time and money.

I especially like this approach, because it acknowledges the reality of business change. Companies usually only change when their biggest customers insist upon it. When large companies recognize they can cut 10%, 20% or more out of their costs by insisting on one-to-one treatment, suppliers are forced to offer personalized services. Once the initial costs are absorbed, this kind of personalized treatment doesn't cost the suppliers more money. In fact, it benefits the suppliers by making it more convenient for their largest customers to remain loyal, and it lessens the degree to which they must sell on price alone.

Thought Exercise ...

What Could You Remember?

Most companies miss numerous opportunities to remember information in a manner that could deliver meaningful benefits to *both* the company and its valuable customers, employees, suppliers, and partners. For example, a catalogue company could simply remember that a customer is also an investor, so that it can provide that person with special services warranted by their willingness to support the company in many ways. When I say remember, I don't mean that the information exists somewhere in the depths of a database, where it can't do the person any good. I mean remembering information and using it in a manner that actually benefits the person. An employer could remember which employees have a desire to work in other countries, or which speak a foreign language, so that they could be notified first when opportunities arise that are related to their interests.

Pick a group of customers, employees, or any other type of stakeholders and use this chart to brainstorm new types of information your business could remember for them. Then, list the ways you would use this information, and rank the likely benefits to your firm and to the stakeholders. Pay close attention to the ones that garner "3" ratings in both columns and ignore those that have high ratings for the company, but not for the person.

What Could You Remember?

Information remembered	Intended use	Estimated benefit (0 = none, 3 = high)	
		To your business	To your stakeholder

Notes

1   "Network Solutions Offers Its Database of Domain Names to Marketing Firms", by Thomas E. Weber, *Wall Street Journal*, 16 February 2001.

2   "Employers can read your email from today," ZDNet UK, 11:45 Tuesday 24 October 2000, Will Knight
   http://news.zdnet.co.uk/story/0,,s2082152,00.html

3   *Workplace Testing: Monitoring and Surveillance; a 2000 AMA Survey.* Figures taken from Summary of Key Findings.

4   *The Surveillance Society: The Erosion of Privacy in America*
    http://news.mpr.org/features/199911/15 newsroom privacy/desktop.html
    Reporter: Bill Caitlin. Minnesota Public Radio.

5   Don Peppers and Martha Rogers identified IDIC as the principal strategies of
    one-to-one marketing, which focuses on providing personalized service to a
    firm's customers. First, *identify* individuals. Next, *differentiate* the way you
    treat each person, based on his or her needs and values. Then, *interact*
    individually with each person to better understand his or her needs.
    Finally, *customize* the services you provide. For more information, read
    their book, *The One to One Fieldbook*, which was co-authored by Bob Dorf.

6   Net Nanny press release dated 18 December 2000.

7   Bruce Tulgan (2001) *Winning the Talent Wars*, p. 156, Nicholas Brealey.

2

Understand the Fears

T om Montgomery, an executive at United Messaging, was telling me a story I'd told hundreds of times before, or so I thought. Turns out he had a different ending in mind.

"We would all like to get back to the old-fashioned service where you return to your local merchant and he remembers that you buy large white eggs and that you like a special kind of fabric. But we wouldn't think so wistfully about this type of relationship if the merchant had run off and shared intimate details of your life with the blacksmith, the saloon owner, and the dressmaker."

There is a growing, gaping disconnect between the way most companies do business and the desire of most people to have control over who gets access to their personal information. Companies want it all. They want to collect personal information, but not limit their ability to use such information to their benefit.

My brother recently went to racing school, where he spent time in an assortment of racing cars. After 24 years of driving experience, he already knew how to drive. Yet, he described going into a turn as an unsettling experience. "The car is so much faster, it's a completely different experience," he said. "You can't look in the rear-view mirror, or you'll miss the turn. Nothing you know works anymore, and you have to learn all over how to drive."

This is what information technology is doing to the average company. It's so much more powerful – and far-reaching – than our old tools that companies must relearn how to interact with their stakeholders, or they will horribly mangle the relationships. But they must also learn that simply because their new car can go 160 mph, that doesn't mean it's safe to drive that fast.

Over the past nine months, Wendy Smith and her team had proven their value. By searching through the vast databases owned by the leading firm, they had been able to uncover eight patterns that each represented a minimum of £125,000 in either expense reductions or incremental revenue.

Wendy described her work at PXR Technologies as unearthing the patterns no one knew existed. Such as the fact that employee behaviour changes six to nine months before he or she quits, which was roughly two to four months before the employee decided to look for a new job. "Wait a minute," the firm's CEO, Jeremy Lucas, had said in their first meeting, "You're telling me you know an employee will start looking for a job before the employee himself knows?"

That was exactly the case. By analysing three years worth of performance reviews, exit interviews, expense reports, email and phone records, business unit results, productivity reports, and internal memos, the system Wendy's team built could spot subtle changes in employee behaviour that preceded a job search. Some employees travelled less, others more. It depended on the nature of their job, whether they were in management or sales (the latter tended to travel more.) Contrary to conventional wisdom, productivity during the pre-decision period often went up, not down. This happened when the trigger for a job search was the company's failure to reward or recognize outstanding performance. "You have many employees," explained Wendy, "who are more talented and aggressive than their immediate superiors. Such bosses often have an involuntary tendency to slow their subordinates' advancement. The result is that we lose good people when they get tired of fighting for what they've earned."

Armed with Wendy's data, the firm had made subtle but effective changes in its HR practices. Some employees were pushed out early, saving the company the cost of having a disgruntled employee spending five months searching for a job on the company's time. Others were promoted, or received bonuses, thanks to the intervention of senior level executives. To be discrete, Jeremy had insisted that Wendy's group – and her data – were kept far removed from this process. Instead, their findings were summarized in several memorandums issued from his office, positioned to look like common-sense initiatives, and nothing more.

But now Wendy was arguing that the firm should take a quantum leap forward. "Our job search predictor deals with problems that were created after a person was hired. We should be intervening much earlier, before a person is hired." She wanted to supplement the group's extensive data analysis with what she called explicit information: data that the company gathered from prospective employees before and during personal interviews. It made perfect sense, but was it legal, or ethical?

The pressures he felt tempered Jeremy's sense of caution. With every passing week, the stock market increased in volatility and punished more severely the slightest misstep or disappointment. Last Tuesday, PXR's stock lost 25% of its value – £200 million – because a competitor announced disappointing earnings. He snapped back to the discussion that had continued while he was lost in thought.

"We don't need to ask any questions that are prohibited by law or convention," Wendy was explaining, *"But we can still gather the type of knowledge that would be ours if such prohibitions didn't exist. There are subtle changes that take place in the way you think when you're considering having a baby, or when you find out that you have a genetic predisposition to a disease. By altering some of the questions in standard personality and basic job-skills tests, we can spot many of these changes."*

"OK, for the sake of argument let's assume that's true, and we don't mind how unbelievably illegal this conversation is," said Geoff, the firm's COO. *"What the hell are we going to do with the results? If we give every manager a scorecard that lets them determine whether this woman is about to have a baby, or that guy is going to loaf off for six months, they'll quit, and we'll get sued big time."*

"That's where the neural network comes in," said Wendy. *"No person will be involved. The HR system itself will decide. We'll program it so that it never states a reason for rejecting a candidate, but simply reports that the prospect's characteristics aren't a good fit with the culture and approach of the firm. Even I won't know whether a specific candidate was rejected because she's a hothead or because she's with child."*

"How much can we save with this system?" Jeremy asked.

"Over £2.5 million next year, taking into account not only lost productivity, termination, and retraining costs but also the cost of bad decisions made by people we never should have hired in the first place."

Jeremy thought about the last board of directors meeting. The message was clear. We don't care what it takes, get this company ahead of the pack.

"OK, do it," said Jeremy. *"But send me a memo that justifies this expense without mentioning any of the borderline stuff, and don't let*

anyone but you, Wendy, program the sensitive elements of your neural network, whatever that means. Don't wait for my official approval to move ahead. We need all the weapons we can muster to hit our targets next year."

Not long ago, an editor from a leading direct-marketing publication called to ask why the media were making so much fuss about privacy. After all, he reminded me, companies have been buying and selling individual information for years: who bought certain items within the last six months, whether they were male or female, where they live, and so on.

Yes, I answered, but companies used to be really bad at collecting and accessing such information. They knew a customer bought dog food, but they didn't know where that customer worked, what store she visited before ours, her exact age, the car she drives, and the ages of her children. Now, companies are getting very good not only at gathering data about individuals, but also at analysing that data to spot patterns than were not obvious before.

When it comes to privacy, I don't fear new companies with innovative business models as much as I do older, more traditional firms. Some of the newer firms may be aggressively collecting personal information, but most have spent a good deal of time and money understanding privacy issues. It takes malevolence for these firms to invade privacy. At traditional firms, managers are more confident in certain business practices that are "commonly accepted", and thus are much slower to recognize that business as usual will result in unintended invasions of personal privacy. All it takes for traditional firms to invade privacy is inertia. The threats to privacy don't come from some distant entity, they come from the way you and I think.

To simplify these complex issues, I'm going to focus on two different types of stakeholder: employees and customers. For the purposes of this chapter, in most circumstances you are either working for a company, or being served by one.

Buying Time? Privacy at Work

When you think about privacy at work, no issue is more controversial than employee monitoring, which is the practice of watching the actions and interactions of employees by monitoring their calls, emails, Web browsing, or other activities. A survey released in January 2001 by KLegal, the law firm associated with KPMG, found that 50% of the companies surveyed admit to monitoring employee use of email and the internet only "infrequently". But 20% of those who monitor usage have done so without their employees' knowledge or consent.

Stephen Levinson, Head of Employment Law (UK) at KLegal, said, "Recent increases in regulation have created uncertainty about how to manage employee use of the Internet and email." He continued, "The 20% of employers who monitor without permission face potentially expensive civil claims as well as possible intervention by the Data Protection Commissioner."[1]

Whether or not you feel comfortable with employee monitoring depends in large part on how you define what a company gets in return for paying an employee. Let's look at two extreme views.

The traditional view is that a company pays an employee to work certain hours and that the company has a right to dictate how the employee behaves during those hours, as long as the company observes all relevant laws. Anyone who punches a time clock is painfully aware of the control exercised by such an employer. I remember one media report on a meat-processing plant that described how employees were prohibited from going to the bathroom except on their breaks and lunch hour. In the traditional view, the company can forbid phone calls, emails, letters, and other tasks of a personal nature. Or, the company can permit such activities, but can monitor calls and emails that go through company-owned or operated systems.

A non-traditional view is that a company is buying a certain outcome, and that within reason it is up to the employee to produce or exceed the expected results. When I worked in television, the station could never find John, the most successful sales representative. He outsold his nearest colleague by a long margin, and although we used to speculate on John's whereabouts (hint: he lived at the beach), management never asked John to account for his time. They didn't care whether he worked 1 hour a week or 60; they just wanted the revenues he produced. This approach gives the employee broad discretion to do whatever he wants, at home or at work. In fact, we welcomed it when John made personal calls from the office; at least we knew where he was.

The traditional view grows out of a traditional economy, one driven by mass production. In this sort of system, profits are a by-product of manufacturing. It's relatively easy to quantify a person's contribution, and most people contribute by working hard, hour after hour, day after day. But economies in the most developed countries are increasingly driven by intellectual capital, the business of ideas. Whether he is writing software, conceptualizing a new business partnership, or inventing a more efficient process, the employee capable of performing such work carries the company's best assets around in his head, and insights don't come only during normal business hours.

Clearly, there are still businesses that must operate in the traditional manner. Here's what characterizes such businesses:

- *Low trust of employees* – Companies that use unskilled or unproven employees, or that are in a cash business, typically act as though they do not trust many of their employees. If they told employees, "You don't have to punch the time clock anymore, just tell the manager how many hours you worked last week," they would expect many employees to report their hours inaccurately.

- *Location and time dependence* – Manufacturing plants, restaurants, shops, and doctors' surgeries all require certain employees to work at certain times, and to be completely focused on their work.
- *Client preferences* – Driving instructors, accountants, and masseuses all sell their time by the hour, largely because clients are comfortable paying for their services in this manner.

Employee-monitoring issues arise when companies require employees to work certain hours *and* give them latitude to manage their own time. The vast majority of managers and professionals fall into this category. They spend long hours at work, and if they are married, their spouses are increasingly likely to work as well. That means neither has time during the week for the countless details involved in managing a household and/or raising kids. They have little alternative but to use their employer's email and phone systems.

There are other circumstances that require employees to perform tasks during "work" hours that they prefer to keep confidential. One is exploring other career options. It has always struck me as hypocritical that many companies simultaneously insist that looking for a job should be done in an employee's own time, and yet ask prospective employees to interview during normal work hours.

In cases of personal or family illness, an employee may prefer privacy. Whether the illness is merely embarrassing or potentially deadly, its implications could impact a person's ability or desire to work, yet the employee deserves the freedom to decide whether this will be the case before informing his or her employer.

There are also times when employees express opinions about other employees – their boss, perhaps – that they do not intend to be received by others in the firm. Is it right that a comment written in haste to a peer ("Do you know what my boss had the audacity to say?") be unearthed by an automated system and shown to the boss?

If employers insist on monitoring employee communications through company systems, they create an environment in which employees are motivated to reduce the hours they work. Even worse, such practices demonstrate that the firm does not trust its employees. So why shouldn't the company limit such practices to new, unproven employees?

Ken Segarnick, assistant general counsel at United Messaging, argues that email monitoring programmes have to include all employees, not just lower-level ones. Otherwise, he says, the programme is ineffective and unfair. "There is a legitimate need to monitor an email system, to safeguard some of the liabilities, and to ensure a safe and friendly work environment."

In other words, companies operate email systems and thus have a duty to ensure they aren't used to support illegal practices such as sexual harassment or discrimination based on race or religion.

I didn't agree with Ken's observation at first, believing that there have always been clear, justifiable differences in the way firms treat senior level executives versus lower level associates. But the issues involved with privacy are so sensitive, personal freedom depends on everyone having the same rights and the same obligations. Sexual harassment, for example, is not just a problem confined to the junior, "less responsible" associates of a firm. On the contrary, the more senior an employee, the more power he or she wields that could potentially be abused.

But there's a far greater issue, which is the need to avoid a Big Brother situation in which a central authority monitors intimidated masses. By including all employees in all employee-monitoring programmes, there will be a natural tendency to minimize monitoring. Let's take the example of using message monitoring to ensure employees don't use obscenities. At first glance, this sounds like a good idea, a way to increase civility and prevent abusive treatment of employees. But, in the real world, people swear, and email is an informal communication medium. In my experience, some of the most senior managers use the worst language. A number of these executives wouldn't tolerate for two minutes a system that edited or rejected their messages because they said, "This is a shitty idea."

If you insist on promoting civility at a firm, a better approach might be to create a "sanity check" function that operates locally on each machine, much like a spellchecker. So when you hit "send" too quickly, a message would pop up and ask, "Do you really want to call the recipient of this message an 'ignorant son of a bitch?'" This would be a far less expensive and less intrusive approach.

If firms feel compelled to protect themselves from their employees, it will pull us into an era of unprecedented employee surveillance, which could extend into every possible means of surveillance. Companies will have a "duty" to monitor every new form of computer-based interactions. Let's take a look at what this duty will produce.

First, a practical point. For efficiency purposes, you don't literally watch every employee. You analyse the data that describes their actions and interactions. Roger Clarke explains, "Watching people is expensive, even when technology is applied to the task. It's far more resource efficient to monitor people through their data. People leave behind them large numbers of data trails, and these can be used as the basis for identifying individuals worthy of closer attention."

Clarke, a prolific author on privacy issues who coined the term "dataveillance," compiled the following summary of the types of surveillance companies can leverage. I've provided the explanations, to make them relevant to our discussion of employee privacy:

- *Visual* – Cameras are everywhere ... in public spaces, in the corridors at work, and mounted in increasing numbers of our computers. Facial-recognition technology is starting to make it possible for automated camera systems to identify a specific person and log their actions in a database. As Clarke points out, this will make it extraordinarily more efficient for firms to monitor the movements of employees.

- *Audio* – Digital audio recorders make it possible for employees to record conversations they have with each other and for business units to record meetings and presentations. In the interest of knowledge sharing, it makes sense for these interactions to be stored on company intranets. As voice-recognition technology finally becomes practical, you'll be able to search for topics, phrases, and ultimately specific voices. Imagine your boss wondering: I wonder what (your name) says when I'm not in the room? He'll be able to search a database and hear for himself.

- *Phone tapping and encryption* – This is generally the domain of governments and law enforcement, but firms often keep records of the phone numbers employees call and the duration of such calls.

- *Voice and pattern recognition* – Computerized systems are getting better at recognizing the meaning of words, certain groups of words (for example, "we have inside information about this stock"), or a common sequence of events. The fictional story that opens this chapter illustrates the growing power of firms to mine and analyse databases, spotting patterns that were never before recognizable. Pattern recognition also underlies biometric measures such as fingerprints, iris and retina scans, and facial recognition.

- *Proximity smart cards* – This type of ID technology gets you into and around secure buildings and enables you to charge meals in a cafeteria or pass through a motorway toll booth without slowing down (that much). Every time you pass a sensor it creates a record of your movements.

- *Transmitter location* – Your mobile phone regularly searches for incoming messages. To do this, it must be recognizable to telecommunications towers. By triangulating these towers – looking at the signal from different directions – your communications provider can identify your location, as long as you have your phone or pager switched on. The leader of the Chechen resistance against the Russians was killed by two ground-to-air missiles that locked onto the signal provided when he used his mobile phone.[2]

- *Email at work* – One of the least understood aspects of email is that you can't delete an email message; you can only remove it from your own records. But one or more copies will still exist on your employer's email system and on those of the people with whom you communicate. Security experts point out that even when you delete a message or file from your hard drive, you are not really deleting the file, but instead deleting the pointer that tells your system where to find the file. If you want confidentiality, don't use email.

- *Electronic databases* – Whether information comes from one of the sources on this list, or from hundreds of other sources, data exists about a person in credit reporting, telephone directory, voter registration, motor vehicle, and other databases. Most of this information can be "rented" by your employer, or by anyone else. Once this data is inside a firm, theoretically any employee can search it, if they have permission. The trend is to allow more employees, not fewer, direct access to databases. Haven't you heard consultants and software companies evangelize, "Unlock the valuable data in your databases?"

- *The Internet* – On *60 Minutes*, Junkbusters founder Jason Catlett observed that people won't tolerate in real life the type of tracking they submit to online. "Suppose every time you walked around the mall, somebody put a bar code on your shoulder ... and scanned your shoulder ... and went to a database, saying, 'Ah, yes, that's Lesley who visited the shop next door fifteen minutes ago.' That's the level of surveillance that's going on on the Internet." Companies track your actions at their site, and across other sites. Unless you take precautions, some companies will place cookies on your computer without your knowledge or permission.

Theoretical discussions about accepted ways to use these monitoring methods understate the chilling impact employee monitoring can have. I know of one dot.com company that was recently battling for its life. Concerned that his senior management team was talking to headhunters, the firm's president asked his MIS (Management Information Systems) manager to provide him with access to outbound employee emails, so that he could determine who was committed and who was not. Uncomfortable with this, the manager went to one of the senior managers, who promptly convened the other executives, minus the president. They then gave an ultimatum to the board of directors: either he goes, or we do. The company eventually folded, and several of the executives cited this as the pivotal moment that destroyed the firm's ability to survive.

The Personalization/Privacy Trade-off

Not all potential privacy issues arise from a practice as overt as employee monitoring. Sometimes, practices that make life easier for employees also force them to accept less privacy.

Remember the story about Fred Jones, who used a knowledge management system to find an engineer in Australia who knew how to solve his problem? Knowledge management systems are not new, but until now most have managed documents. Most knowledge still exists in the brains of people rather than transcribed in a document. New approaches are coming to market from firms including Autonomy, Tacit, and Net Perceptions that include people in the knowledge management framework. To accomplish this, they need to know what you know, and often what you are working on.

There are a variety of approaches to acquiring information about what you know. Some systems use email, suggesting additional recipients who would like to be copied on a particular message. (One partner of mine

More Data Trails Everyday			
Old trails	Recent trails	New trails	Developing trails

Credit-card transactions
Other charge accounts
Checking accounts
Loans
Taxes
Licenses
Catalogue shipments
	Cash-dispenser
	  transactions
	Debit-card payments
	Telephone-call record
	Loyalty cards
	Frequent flyer numbers
	Voice mail
		Building-access security
		Video surveillance
		Mobile-phone records
		Mobile-phone locators
		Pagers and wireless
		  PDAs
		Email traffic
		Web traffic
		Store value cards
		Caller ID
		Electronic-toll
		  payments
		GPS devices for cars
			Personal telephone
			  numbers
			Fingerprint readers
			Retinal and iris
			  scanners
			Facial recognition
			Voice recognition
			Keyboard (typing)
			  recognition
			"Black boxes" for cars

Adapted from Roger Clarke's *Trails in the Sand*, 18 May 1996 http://www.anu.edu.au/people/Roger.Clarke/DV/Trails.html			

would be outraged at this obvious lack of respect for what he calls "the electron conservation society". People who get a hundred emails a day don't want more.) Other systems create profiles of users and their documents, so that they can identify a particular expert in the pharmaceutical applications of wild brush in the Urals when someone in your firm needs one.

The hardest part about managing knowledge is getting people to reveal what they know, or even what they like. Increasingly we see on the Web and in intranets little checkboxes next to an article or link that says, "I like it." This is a device to get feedback from people about what information you find most valuable. By matching your characteristics and opinions with those of other employees, a knowledge-management system could theoretically eliminate vast amounts of irrelevant information and bring you directly to information you didn't know existed, but find tremendously valuable. Imagine if instead of seeing "Here are the 28,214 references on security," you were given, "Here are the three articles that MIS managers like you in mid-sized manufacturing companies located in the United Kingdom found most useful."

Here's the trade-off: To get significant value, you have to share information about yourself. The more information you reveal, the more likely that the system will be able to tailor its services to your needs. You can choose to be discrete, but doing so may put you at a distinct disadvantage as systems such as these proliferate and your peers – and your competitors – start to use them.

We will also start to experience instances where "opting out" on a firm-wide or even business-unit practice casts suspicion on an employee. Why won't Jim tell us what he's working on? Maybe he really doesn't do anything all day? Did he really go to the meeting . . . if so, where is the transcript he should have posted on the system?

Horrendous Data

The biggest cause for concern, as companies come to rely increasingly on data about employees as well as customers to drive the way they treat each person, is mistakes. The quality of data in most companies is horrendous. If history is any guide, it will take much longer to fix the quality of corporate data than it does to implement the new technologies we are reviewing. That means we are entering a period not unlike the Wild West in America, which I interpret to have been: Everyone has guns, and very few know how to use them responsibly. (Yes, I know some will argue not much has changed.)

It's dangerous, and innocent people will get hurt. It's a lot easier to cause harm with a mistake than to fix it with an apology.

Jane knew she had dead wood in her group, but she had difficulty separating her personal biases from objective decisions about individual performance. Tonight, she was working late to see if she could gather more information about a few subordinates whom she considered borderline.

Tom Douglas was one of the toughest calls. He had been warned twice that his work was below par, although Jane had heard recently that Tom was turning things around. But Jane didn't like the guy. He had a tendency to be a know-it-all guy and just never developed the work ethic the firm needed. Besides, he probably had enough family money to keep him living in style without a job. Being laid off could make his day, for all Jane knew.

Jane decided to review the transcripts of recent meetings to see what Tom had contributed. Over the past six months, every important internal meeting had been recorded, digitized, and converted for storage in a searchable database. She could choose many ways to search, but instead of searching by Tom's first and last name, Jane decided to look at several critical meetings that she knew Tom had attended.

Fifteen minutes later, Jane signed off and started typing a termination notice. It was shocking. Tom barely said anything at meetings he was supposed to have led, and when he did comment, his remarks were unbelievably ignorant.

Two weeks later, Tom was out of a job, which was especially unfortunate, because his wife was pregnant with their third child and contrary to Jane's impression, he neither came from money nor had much saved. But that's not the bad part. There was an error in the transcripts that Jane had reviewed. Tom's remarks were attributed to Tim Driscoll, and vice versa. The reality was that Tom had turned his performance around and was a real asset to the firm. But that's not what the database said.

In the above story, the error was confined to a business unit's database, yet it still harmed someone's career unnecessarily. By all means, Jane should have exercised more care to check her facts, and in a perfect world, Tom would have questioned the reasons for his dismissal and been able to correct the mistake.

This story leads us to two principles that are absolutely critical in managing issues of privacy:

1. Individuals must be able to easily fix mistakes.

This means that individuals must have easy access to all information that a firm stores about them, and that there must be an interface and a process that enables them to correct any mistakes they spot.

2. The magnitude of a database error is in direct proportion to the number of people who have access to it and the implied veracity of the database itself.

If an office of ten people maintains a database and it contains an error, once spotted it is easy to fix. But if 40,000 other people at the firm share that database, it's much harder to know how many people have already viewed the incorrect information or duplicated the error in other databases. Once data spreads in this manner, the original owner loses control of it, unless the data element is simply a link to information held in the original database.

When Duncan Phenix tore open his financial statement early in 2001, he found his boss's information instead, complete with Social Security number, fund balances and birth date. He was one of numerous employees at Raycom Media, which employs 2,800 people, who received statements that combined records for more than one person or that included the wrong person's information.[3]

"There's the saying: 'To err is human, but to really foul things up takes a computer.' Computers amplify simple mistakes," Richard Smith of the Privacy Foundation, a national research and education group, told the *Washington Post*.

Errors get especially out of hand when the database is viewed as an absolute authority. Social Security records, Interpol most-wanted lists, and bank records are supposed to be correct, and it is much more difficult to challenge successfully an entry in one than it is to fix a database maintained by your friends and peers.

The second principle presents a powerful argument that firms should never rent or sell information about an employee, customer, or any other stakeholder. The potential harm to a stakeholder grows exponentially as you allow data to be more widely distributed. We'll talk more about this later in this chapter, after we look more closely at privacy and customers.

There is good news and bad news about data quality. The good news is that companies are pouring money into fixing the quality of their data and connecting databases with each other, so that they can get a complete picture of their interactions with each person. The bad news is that this will erase one of the primary defences of individual privacy, which is the reality that most companies have been too disorganized to build full profiles of many individuals. I recently spent an afternoon with marketing executives from a large, conservative life insurance company. A vice-president asked me about the growing media coverage of privacy. I reminded him about the reason for our meeting, which was to explore the implications of his firm's not only connecting its databases, but potentially sharing policyholder information with other companies. The firm had never before shared such data with its agents, never mind outside companies. If even conservative, slow-moving companies like this one (their characterization, not mine) were considering actions like this, doesn't that justify the "sudden" interest in privacy? The executives agreed that it did.

Tracking and Targeting Customers

One of my pet hates is calling customer service for help with my credit-card account. Inevitably, the recorded voice says something like, "To better serve you, please use the telephone keypad to enter your 16 digit credit card number." You then wait on hold for 45 seconds, when a live person comes on the line and asks, "Can I please have your credit-card number?"

The most plausible explanation for this maddening disconnect is that large credit-card operations have numerous customer-service centres, each serving different accounts. When you type in your account number, that information is used to automatically route your call to the proper centre. Then, you wait on hold until a representative is free in that centre. The problem is that while it's easy to forward a call to one of five centres, it's hard to attach your credit-card number to that call and have your number automatically pop up on the representative's computer screen when he or she answers your call.

Yesterday – honest – I called my bank fully prepared to experience this frustration again, but instead discovered that Citibank has finally overcome the technical challenges. The representative knew my name and account number as soon as he came on the line. I was delighted.

In my experience of talking to people about privacy, people are vastly more upset about how much companies forget about them than they are upset about what companies remember.

Paradoxically, the biggest threat to customer privacy comes not from the companies with whom we do the most business, but from those who are obsessed with selling us something that we don't want to buy. The story about tracking consumers in Chapter 1 illustrates that companies have an easier time buying information about prospects than they do remembering knowledge about customers. In the former case, they can rely on the information-gathering capabilities of database marketers and need only to be able to pay the cost of acquiring such information. In the former, they must first identify each customer – a harder task than it sounds for many firms – and then develop a process to gather and store personal information. This is hard, especially since most customers lack a motivation to share such information in the absence of immediate and significant rewards.

As a result, the vast majority of companies continue to take the easier path, which is to buy personal information and use it to target and track potential customers. In this situation, companies have little to lose. They don't have a relationship with the targeted people, so the worst that can happen is no progress, right? No. In reality, the worst that could happen is that the company's name gets plastered across every news front page and the firm's market capitalization drops 40% in a single day.

Several entrenched factors make it hard to believe this situation will change without a concerted effort by a firm's senior management and board of directors. First, the vast majority of managers are paid to sell products, not serve customers. I've worked with many rapidly growing software companies and know not to bother them during the last weeks of any fiscal quarter, because everyone is focused on closing deals. In many cases, deals get closed that aren't good for either party, but the compensation system rewards revenues, not profitable sales. As long as compensation systems work this way, firms will gather whatever information is legally available that helps them close deals.

Second, most firms are organized around products, instead of around customers. This means that managers are paid to manage and sell a limited number of products, which gives them little or no ability to react in a meaningful manner to feedback from customers. More importantly, they cannot meet their goals by broadening a customer relationship – building share of customer, as Peppers and Rogers call it – but only by finding new customers who want to buy the relatively few products they sell. This reinforces a system that values prospects more highly than customers and that makes personal information a valuable tool in hunting down and acquiring new customers.

Third, many of today's marketing executives have a background in direct marketing, which is basically a discipline designed to generate an acceptable number of orders from an accumulation of solicitations. In direct marketing, it can be a great success if 99 out of 100 people consider your mailing to be a worthless, annoying intrusion, as long as one person orders at or above the targeted level. That's because one order can be enough to justify the cost of sending out one hundred solicitations.

Until now, most consumers considered direct marketing to be nothing worse than an annoyance, because it has been relatively easy to eliminate the annoyance by throwing junk mail in the trash. But this mindset becomes dangerous when so much data exists, enough to chronicle our movements on an hourly basis. Ten years ago, a typical direct marketer would rent a mailing list that told her these consumers spent over £50 in the past year at Marks & Spencers or in Freeman's catalogue. If the marketer wanted to pay a few pounds extra, she could select buyers from certain regions, or of just one sex. (Marketers often conclude that women are more likely to respond to a certain offer than men, or vice versa.)

A direct marketer would pay these extra pounds only if she believed doing so would increase the bottom line; if she had some reason to believe that certain regions would respond better than others – results from a previous mailing, perhaps – she would invest the extra sum.

For direct marketers, it all comes down to a spreadsheet. The goal is to generate a profit, and most will buy as much data as they can if it increases the profit they make. If more useful data becomes available, they will buy it. (Technically, they rent it for a one-time use.)

Armed with a basic understanding of human nature and having felt the pressures that exist on all of us who work for companies, I find it difficult to believe that any direct marketer will pass up a legal, ethical source of data. Which means that they will embrace every new technology that provides profitable information about prospects and customers. They are compensated to use this information to aggressively target any one-in-a-hundred prospect.

On a number of occasions I have conducted one-to-one marketing workshops and, to generalize, the direct marketers have the hardest time accepting the principles that power such relationships. Relationship is a fuzzy term. They can't quantify the value of a relationship, but they can of a transaction. When you are used to ignoring 99 people to get an order from one, it's hard to make the mind shift to start thinking about the nature of your interaction with the other 99, or about their right to privacy.

Taken together, these entrenched factors produce a situation that completely lacks the win–win characteristics so critical not only for one-to-one relationships, but also for protection of individual privacy. Through their compensation systems, organizational structure, and longstanding practices, most companies are positioned to invade privacy, whether they intend to or not.

You Can't Approve What You Don't Understand

So why aren't consumers even more upset about privacy than the media portrays them to be? And why did I say people are more frustrated by what companies forget than what they remember?

Most people have no significant understanding about the means of surveillance and the data-gathering capabilities we have been discussing. It's not their job to look 6, 12, and 18 months down the road, as we are doing. Most people are time starved, more concerned with family, friends, work, community, and life than with understanding the advances, and implications, of potentially intrusive technologies.

In August 2000, only 43% of American Internet users knew that creating cookies is the way companies track Internet activities, and only 10% of all Internet users had set their browsers to reject cookies.[4] This implies that most people don't know that sites place cookies on their PCs without their permission or knowledge, and that 90% of the public is vulnerable to this approach. Despite this lack of knowledge, 84% of these users answered yes when asked if they were concerned about "businesses and people you don't know getting personal information about you and your family." Imagine how high the concern will be when the public discovers the true – and expanding – extent of personal surveillance.

A January 2001 report by Consumers International, a conglomerate of worldwide consumer organizations that includes UK groups such as the National Consumer Council and the Consumers Association, found that despite strict European data protections regulations, UK sites are no better than US sites at informing users how they use their data. Internet.com reported that, "Some of the best privacy policies were found on the US sites."[5]

Most people don't realize that every time they go online, firms such as DoubleClick and Engage are tracking their surfing habits and storing this information in vast databases. This enables firms to "profile" individuals and facilitate other companies to target them for advertising and promotions. Theoretically, these profiles are anonymous, but they don't have to be. People also don't know that phone companies could, if they wished, draw a map of an individual's travels throughout a day, week, month, or year. They don't realize that without greater legal protection than exists today, your boss could notice that your car is parked for two hours every Friday afternoon in the parking lot of a bar. Or that your car and that of a co-worker of the opposite sex are in the parking lot of a hotel ... where no conference is scheduled, and none would ever be.

GeoSpatial Technologies, based in Santa Ana, California, plans to begin selling in the fall of 2001, a vehicle-tracking device, called GlobalTrax. Via a Web browser, parents will be able to monitor the speed, direction, and location of the car their teenager is driving, or adults could do the same for their elderly parents. Laidlaw, the world's largest bus company, is already testing the device as a system to ensure that drivers stop at all railroad crossings and open the doors to check for oncoming trains, as American law requires.[6]

The Institute of Transport Studies at Leeds University has won funding for expanded trials from the Department of Transport, Environment and the Regions. They are testing a system that places speed limiters on cars and could be used to force drivers to stay under the speed limit in certain geographic areas. For example, such a system could prevent cars from traveling too fast in front of a school or playground.[7]

Without an abundance of extra time and an exceptionally creative mind, it's difficult to conceive all the new ways that data can be used. In the north-eastern United States, there is an electronic toll payment system called EZPass, which has made life much easier for commuters by eliminating the need for drivers to stop and pay a toll. When you pass near a tollbooth, it recognizes the electronic pass mounted on your front windshield and automatically charges the toll to your credit card account.

On a monthly basis, you get a listing of when and where you passed each tollbooth. In a situation such as exists on the Massachusetts Turnpike, where there are tollbooths at both the entrance to and exit from the highway, the EZPass system could be used to provide perfect enforcement of speeding laws. If you arrive in fewer minutes than the speed limit allows, a $125 penalty could be automatically charged to your credit card. While people would be outraged by this, it makes little sense that today EZPass users still engage in a game of cat-and-mouse with state police, while proof exists which ones were speeding.

Privacy advocates, who spend a great deal of time thinking about these issues and where technology is leading us, worry that we are moving towards a society in which every individual can be identified perfectly, all the time. Simon Davies, director of Privacy International, said, "The core lesson from history is that perfect identity always equates to perfect control. Every free and open society in history has always had a degree of anonymity. It might excite organizations to think that we could live in a better society if everyone was perfectly identified. But we think that it's a nightmare."[8]

Davies also reinforced a point we discussed earlier, which is that it's easier to be comfortable with having your data in a small, closed system than in a database open to the world. He pointed out that this tendency presents a significant stumbling block for biometric systems that identify characteristics such as your fingerprints, voice, and iris patterns. You might be comfortable allowing your bank to use your iris pattern for the sole purpose of identifying you at their bank, but do you really want the bank to include your iris pattern in a global database that can be accessed by thousands of companies and hundreds of governments?

People don't think about the weaknesses of biometric systems, such as the fact that many can be fooled. Holding up a photograph of a face can fool some facial recognition systems, because they lack the ability to test for "liveness". (One way to test for liveness in such a situation is by detecting the warmth associated with blood flowing under a person's skin, instead of just looking at the shape of a face.)

These systems use digitized information, so someone could steal a digitized file that represents your fingerprint, and then transmit that file to the biometrics system to fool it. In this situation, the thief doesn't actually need your fingerprint, because the system doesn't have it either; it has a copy of your fingerprint file, too.

Janlori Goldman, co-founder of the Center for Democracy and Technology, observed, "Once divulged, bits of personal information can reveal what we think, believe and feel. Personal information, disclosed over a period of time in a variety of circumstances, can be culled to create a 'womb-to-tomb' dossier. As people lose the ability to control how others see them, and judgements are made about them based on information gathered third-hand, people grow to distrust information-gathering entities."[9]

For a small taste of this, try visiting Deja.com, (now owned by Google.com), the largest archive of discussions on the Net. You can search for all the postings by a particular author, and if you have sufficient time, can develop a far deeper sense of a person than you would get from reading a single message. One person I chose at random had posted 580 unique messages. He didn't necessarily intend to have all his postings displayed at one time, just as you might not feel comfortable with having every purchase you have made over the past three years posted on the Web – or on a private database – for all to review.

Criminals, Sociopaths, and Hackers

So far, I've made the assumption that companies and the people within them are law-abiding and reasonably ethical. If they make mistakes, it's due simply to structural flaws in their organizations and processes. But, as anyone who watches even three minutes of television knows, criminals, sociopaths, and hackers also populate the world.

To the smartest of these populations, vast databases of sensitive information are like red flags to a bull. Isn't it thoughtful, they might think, for leading companies to gift-wrap so many treasures for them?

I'm not a security expert, but privacy and security go hand in hand, so I've taken pains to understand the reasons why we should feel confident that our personal information is safe. It seems there are none. No kidding.

The two best ways to prevent data theft are:

1. don't collect information in the first place; and
2. encrypt the data in your files so that even if someone steals it, he can't use it.

The first point isn't as silly as it seems; for security as well as expense reasons, companies need to think harder about what information they really need to enable meaningful personalization.

No matter how you look at business systems, they are becoming increasingly vulnerable from a security standpoint. We'll talk more about this in Chapter 5, but the more systems are networked together and the greater the use of modular capabilities, the more potential security flaws. By modular capabilities, I mean that many software applications you consider to be a single program are actually comprised of multiple apps that seem to operate as a single system. Someone seeking unauthorized access to a system could exploit each of the "doorways" between these components. These doorways allow modular applications to pass information back and forth to each other, which in a highly simplified example happens like this:

SHIPPING MODULE: Has customer 564789 been authorized for shipment?
BILLING MODULE: Yes

SHIPPING MODULE: What is the shipping address?
BILLING MODULE: 2 Main Street

As an individual user, the best you can hope for is that a large company with a reputation to protect will assume liability for its security lapses, when they happen, which will be with increasing frequency in the months and years ahead.

Of course, this won't protect you if someone decides not to attack an entire database, but rather to steal your identity, or at least borrow it for a few months, leaving you to cope with the consequences for the next decade or two. Darlene Alexander was turned down for a $75,000 mortgage because a credit check revealed that she had too much debt, which included a $22,800 loan for a Chevrolet Camaro. In reality, Alexander was the victim of "credit doctors," who steal clean credit histories and sell them to people with lousy ones. She had no debt, and owned a 1983 Datsun, which she had long since paid off.[10]

In the spring of 2000, California Public Interest Research Group (CALPIRG) conducted in-depth interviews with 66 victims of identity theft, finding that 55% of the victims consider their cases to still be unresolved, an average of 44 months after the ordeal began. Victims reported an average of $18,000 in fraudulent charges, with the highest charges reaching $200,000. The report says, "A victim from California felt that resolving her problem was 'nearly a full-time job'. Robin, a victim from Los Angeles, explains, 'One bill – just ONE BILL – can take 6–8 hours to clear up after calling the 800 numbers, waiting on hold, and dealing with ignorant customer representatives.'"[11]

The less people have to lose, the more they have to gain by going after a company and the personal information it possesses. A 1999 report from the Rand Corporation warned, The rise of networks is likely to reshape terrorism in the Information Age and lead to the adoption of netwar – a kind of Information Age conflict that will be waged principally by non-state actors. There is a new generation of radicals and activists who are just beginning to create Information Age ideologies. New kinds of actors, such as anarchistic and nihilistic leagues of computer-hacking 'cyboteurs' may also partake of netwar."[12]

Not surprisingly, cyberspace hacking wars are breaking out where global tensions are highest. During the fall of 2000, when Israeli/Palestinian animosities intensified, *Wired News* reported that pro-Palestinian hackers defaced at least 40 Israeli sites while Israeli antagonists marred 15 Palestinian sites. "We expect to see more wars like this one being waged out there," said James Adams, chairman and CEO of iDefense, an international private intelligence firm. *Wired News* said Adams foresaw "the future of warfare as one conducted not only by nations with armies, but by individuals with common gripes, banning together against a common enemy."[13]

There are no safe paths in life, and no absolute protection from evil. But it makes sense that the more a company is focused only on its own goals, the more exposed it will be to deliberate attacks.

Privacy issues are growing every day, structural issues prevent or slow companies from responding, and security issues threaten large databases of personal information. What are the odds that governments will sit on the sidelines and allow companies to self-regulate their actions with regards to privacy and personalization? Zero. But is this good news? Do we really want to rely on governments – who brought us taxes, bureaucracy, and spy agencies – to protect our privacy?

Thought Exercise ...

Where Do You Draw the Line?

You may not want to write down the answers to this exercise. Forget about your job for a moment. How much privacy do you expect for yourself, and for those people close to you? The goal of this exercise is for you to come up with four or five places in which you would draw a line that should never be crossed by any business. Perhaps you don't want your employer to remember the intelligence level of your children, or your insurance company to know your predisposition to life-threatening diseases. You might not be comfortable with hotels remembering after you depart which phone numbers you called. Some people don't want companies to keep their credit cards on file. Everyone is different. Where do *you* draw the line?

Notes

1 "The Uneasy World of E," KLegal Internet Survey, 19 January 2001
   http://kpmg.co.uk/kpmg/uk/press/detail.cfm?pr=837

2 http://screenmedia.com/Excursus/Articles/eye.htm
   Dataveillance, 12 December 2000.

3 Ariana Eunjung Cha, *Washington Post*, 24 January 2001.

4 *Trust and Privacy Online: Why Americans Want to Rewrite the Rules*, The Pew Internet and American Life Project, 20 August 2000.

5 "Web Sites Still Fail to Protect Customer Data," Internet.com, 29 January 2001, by uk.internet.com staff
   http://internetnews.com/intl-news/article/0,,6_572581,00.html

6   "Eye in the Sky Lets Net Track Cars", by Earle Eldgridge, *USA Today*, 20 December 2000.

7   " 'Promising' tests on car speed limiters extended," Martin Wainwright, The *Guardian*, 20 January 2001
    http://guardianunlimited.co.uk/uk_news/story/0,3604,425344,00.html

8   Interview published in the *Biometrics in Human Services User Group Newsletter* on 1 November 1998 [Volume 2, Issue 5]. This is an incredible resource. For more information, visit
    http://dss.state.ct.us/digital/faq/dihsug.htm

9   "Privacy and Individual Empowerment," an essay in *Visions of Privacy: Policy Choices for the Digital Age*, edited by Colin J. Bennett and Rebecca Grant, North York, Ontario: University of Toronto Press, 1998, p. 102.

10 *Computer Ethics: Cautionary Tales and Ethical Dilemmas in Computing*, by Tom Forester and Perry Morrison, Cambridge, Mass.: MIT Press, 1994, p. 132.

11 http://privacycouncil.com/linksold.htm?
    http://pirg.org/calpirg/consumer/privacy/idtheft2000/,
    links_identity.htm

12 CNN online
    http://cnn.com/TECH/computing/9904/26/netwar.idg/

13 "Hacker War Rages in Holy Land," by Carmen J. Gentile, *Wired News*, 8 November 2000.
    http://wired.com/news/print/0,1294,40030,00.html

3

Anticipate New Laws

O ne thing is certain: technological advances will force changes in the laws around the globe that protect individual privacy. If you wait for these changes to become obvious, you will forfeit a powerful competitive advantage. People trust leaders, not followers. Once legislation creates new standards for appropriate behaviour, the public will be drawn to companies that can claim to have followed such standards before they were mandatory.

Each country, and in some cases each region, is taking a different approach to privacy, and it's beyond the scope of this book to chronicle them. I just want you to have a taste for the forces and counter-forces at play in this debate, and to provide that, let's start with one of the more entertaining interchanges that took place in the US Senate last year.

Feel free to skim the excerpt below, which represents a tiny glimpse into the type of hearings that precede new legislation. The passage can be confusing and hard to follow, but that epitomizes the legislative process, especially in the complex areas of technology and privacy. To make it a bit easier to understand, I've included my own comments in italics. (I did not participate in the actual hearing.) The players in this interchange are:

- The politician – Senator John McCain of Arizona.
- The corporate representative – Jill Lesser, vice-president of domestic public policy, for America Online.
- The industry representative – Christine Varney, senior partner of Hogan and Hartson, representing the Online Privacy Alliance, a cross-industry coalition of more than 80 global companies and associations that includes AOL, DoubleClick, Acxiom, Compaq, Equifax, and Dell.
- The privacy activist – Jason Catlett, the outspoken president and founder of Junkbusters, a privacy advocacy firm that helps people get rid of junk messages of all kinds: spam (irrelevant messages sent on the Internet), telemarketing calls, unwanted junk mail, and junk faxes.

This testimony was given before the Senate Committee on Commerce, Science, and Transportation on 25 May 2000[1]:

Senator McCain   So, Mr Catlett, along those lines, I, like many others, buy books online, and now when I go on one of these websites, they say, "Hi, John; we just got in a new biography of Napoleon we know you would like," which is true. They know what my preferences are. So actually they're helping me by informing me of books that I would like to read. What's wrong with that?

Catlett   It's a wonderful service, sir, and I use it myself.

McCain   Then you know what I'm getting at here, OK? Where does the line stop where they're informing me and helping me, and they're invading my privacy?

Catlett   Everybody wants the benefits of personalized technologies, and the Internet is wonderful at providing that, provided that the personal information is treated fairly. And that means several things. Only using the information for the purpose that they collected it for, in the case of, say, making book recommendations, and for not selling to or giving to journalists who want to get a psychographic profile of the individual who buys the books. Secondly, the individual should have access to that complete profile that's built up so they can be sure for themselves ...

McCain   Like the FOIA. Like a FOIA, the Freedom of Information Act. (*Washington, like any political capital, has a language all its own. Unfortunately, this makes it more challenging for the rest of us to participate. The senator seems to realize that and is translating for us.*)

Catlett   Precisely, sir, and those laws should apply very broadly to all commercial entities that maintain personal information. It's the right of people to determine the information that's held about them. That information is being used by companies supposedly for their benefit, and so people have the right to see that information.

McCain   Do they now?

Catlett   No, they do not, sir. You have the right to see your credit report, but you do not have the right to see the vastly greater profiles about you that marketing companies have.

McCain   Is that fair, Ms. Lesser?

Lesser   I think it's a fair articulation of the current law. I don't think it's necessarily a fair articulation of all business practices. So, for example ...

McCain   Now, wait a minute. Is it fair for me not to know what ...

Lesser   Oh, I'm sorry; I misunderstood your question.

McCain   Amazon.com's profile of me is?

Lesser   I imagine that if Amazon.com is creating – is giving you, for example, as we do, an opportunity to have a member profile ...

McCain   Is it fair for me to know what the profile is, Ms. Lesser?

Lesser   Sure, absolutely. It is fair for you to know.

McCain   But right now I don't have that right.

Lesser   You will probably be given a right to know what your profile says by a lot of companies, because it's smart business practice.

McCain   But if they don't choose to ...

Lesser   Now, the level of – there's a difference between understanding access, that is, do you access directly into the database or do you have an ability to basically say . . .

McCain   You're complicating the issue. (*Trying to get a straight answer, Senator McCain decides to address his questions to the other panelist. He won't have much success.*)

McCain   Ms Varney, do I have the right to know what profile is compiled on me by an Internet corporation?

Varney   Do I get to ask you a question back to further the . . .

McCain   Yes.

Varney   OK, thank you.

McCain   Tragically, yes. (*Laughter*)

Varney   Do you want to know – a company is going to take what you've purchased on their website to develop their profile. Do you want access to everything that you've purchased?

McCain   No, what their profile of me is.

Varney   OK. So, you don't care about getting access to your past purchases. You want to see what they do with that information.

McCain   I want to know what the profile is, because obviously they are letting other people know that profile.

Varney   Why are they letting other people know the profile?

McCain   I don't know why.

Varney   What if they don't?

McCain   For profit and fun. (*Laughter*) (*This is starting to sound like an Abbott and Costello comedy routine, but with much higher stakes.*)

Varney   Not yours, Senator, I can assure you. If they're not sharing the profile, does that matter to your question? Because here's what the . . .

McCain   Even if they're not sharing the profile. The FBI has a file on me, and I hope they're not sharing it. Yet, I have the ability – well, I don't really care. (*Laughter*) Most citizens would not want that. So, through the Freedom of Information Act then I can find out – I can get my FBI file. Shouldn't I be able to, through some kind of Freedom of Information Act, know the profile that is kept on me?

Varney   Having been through the Senate confirmation process, I do have an FBI file, and I have reviewed it, and what is in my FBI file are facts and summaries of conversations.

McCain   Should every American have the same right as they do with the FBI file?

Varney   But, Senator, that's what I'm getting at. What's in the FBI file – if the FBI has a psychographic profile on me, I have not seen it. I cannot see it.

McCain   They may and they may not. I've seen all kinds of FBI files.

Varney   Can you see what they have on me?

McCain   You are evading my question. Should they have the right to know the profile that is – should I have the right to know the profile that is kept on me?

Varney   Senator, I don't mean to be evasive. I'm trying to draw …

McCain   So, you're not going to give me an answer. (*Laughter*)

Varney   I am going to give you an answer. I'm trying to draw a distinction …

McCain   If you want to ask me a question, you've got to give me a yes or no answer.

Varney   I will, I will. You don't let me, though. I'm trying to draw a distinction between the data that is used by a company to create a profile. Obviously, you have a right to all of the data, the transactional data. What some of the companies will say back to you, whether or not you accept this argument, is we spend a lot of time and a lot of money and hire a lot of people and do algorithms and all kinds of things to come up with what we think is the profile. It's our proprietary property. Is it good business sense to share it with you? Sure. Do you want to legislate it? Talk to the companies that do it; I don't know.

McCain   So, your answer is I don't know. Now, what's your question for me?

Varney   I asked the question, whether you wanted access to the underlying data or to the profile that the data was used to generate. (*Thanks for reading this far, and now Senator McCain expresses a viewpoint likely to power new privacy legislation . . .*)

McCain   I think I should have access – very frankly, I think I should have access to any information that is collected about me and conclusions that are drawn about me. I think that's the right of citizens, and I don't understand how it could be . . . (*Interjection by other participant, and the testimony took another turn . . .*)

What motivates each of these parties? For Senator McCain and politicians around the world, it is the sense that their constituents are increasingly concerned about privacy, because attempts to monitor and track individuals are proliferating by the day. In fact, monitoring hit the big time on 28 January 2001, when the Tampa Bay Sports Authority used FaceTrac technology to monitor every single person who entered the stadium during Super Bowl week. Using cameras to recognize human faces, the system constantly compared each face to a database of known felons, terrorists, and con artists.[2] The American Civil Liberties Union objected that this practice forced innocent fans into a "computerized police line-up" and called the event the "Snooper Bowl."[3] The *Washington Post* reported:

Cables instantly carried the images to computers, which spent less than a second comparing them with thousands of digital portraits of known criminals and suspected terrorists.

In a control booth deep inside the stadium, police watched and waited for a match.

The extraordinary test of technology during the highest-profile US sporting event of the year yielded one hit, a ticket scalper [tout] who vanished into the crowd, reported an official at the company that installed the cameras.[4]

"This was just the latest tool," said Tampa police spokesman Joe Durkin, who reported that the system made 19 matches during Super Bowl week. All had criminal histories but had committed no crimes of a "significant nature". He said police made no arrests as a result of the surveillance cameras' use.

A *Washington Post* reader, writing on the newspaper's Web discussion board, commented, "It's sad that we have come to this, but it is becoming necessary in an age when the person next to you may pull a gun and blow you away. The person on the other side might be a person wanting to be a martyr for some political cause and have a bomb planted in his clothing. I don't like the idea of 'big brother' tactics, but since I am not a criminal, I have nothing to hide. I don't carry weapons or threaten other people. I don't belong to a terrorist organization or any other group that has strong feelings about the need to overthrow the government. Go ahead. Catch the bad guys. It makes me feel safer."[5]

Another reader disagreed, writing, "The problem will get worse as new technology is brought to bear in situations of everyday life. Sometimes too much knowledge can be dangerous."[6]

This is a positive use of surveillance technology, so you could argue it is surprising that anyone besides terrorists, criminals, and con artists would object. No attempt was made to monitor or control behaviour at the Super Bowl, only to prevent entry by people who were highly likely to be there with malicious intent. But what if this same technology was used to identify every person who:

● drank excessively;
● shouted obscenities;

- moved into a better seat when others left the stadium;
- littered.

Privacy International reports that "In Britain between £150 and £300 million per year is now spent on a surveillance industry involving an estimated 300,000 cameras covering shopping areas, housing estates, car parks and public facilities in a great many towns and cities."

For example, closed-circuit television cameras have been in operation in English football grounds for years. Until now, however, such cameras haven't been connected to facial-recognition systems just now becoming available. These systems access a photographic database of known offenders, searching for matches between people in the database and people attending the sporting event. Eventually, such databases may contain photographs of every citizen. With better databases, authorities may soon be able to automatically deduct say, a £200 fine from your bank account, and schedule 50 hours of community service; you'd get an email notifying you of these sanctions as a result of your (public nuisance) on (14 September).

Enlightened politicians recognize that the use of technology such as this will follow the same pattern that happens nearly every time companies implement new technology. First, the companies use it to automate an existing practice, such as providing security at a public event. But later, as companies come to better understand the capabilities of the new technology, they invent new uses for it. This is what troubles society: How do we ensure that the new uses of surveillance, databases, and interactive technologies benefit individuals, rather than oppress them?

Politicians have to address this concern. The challenge, however, is to identify the minimal level of legislation necessary to protect individuals. Companies influence the proportion of such legislation through their actions and attitudes. If companies view their interests as being unrelated to those of individuals, then such legislation must be extensive and ever expanding. If, on the other hand, companies recognize and continually demonstrate that their needs are interrelated with those of individuals, then far less legislation will be needed. Regrettably, most companies behave today as though the former is true.

The truth is that most established **companies** want as little change as possible. In recent years, most lobbying efforts amounted to pleas that legislators allow industries to self-regulate or attempts to demonstrate how effective increased disclosure efforts have become. Unfortunately, disclosure isn't comforting to people who disagree with the basic practice being disclosed. If you are against employee monitoring, you won't be satisfied just to be told that you are being watched. Likewise, if the burden of action remains on the individual, rather than the corporation, disclosure may not be enough to enable a person to successfully control his or her personal data. This is the basic debate between opt-out versus opt-in policies, which we'll discuss later in this chapter.

The above testimony highlighted three significant concerns that companies have:

Lack of flexibility

One loud and clear lesson of this interchange is that companies don't want to have their hands tied, and they don't want to give up ownership of the data in their files. Lesser and Varney come off as evasive and uncomfortable as they try to avoid the senator's attempts to get the two to agree that individuals have the right to know every piece of information that a company stores about them. This is with good reason. They represent a wide range of different businesses that profit in many different ways. The data they collect includes your purchases, returns, browsing activities at their website, browsing activities elsewhere on the Web, emails you have sent or received, segments into which they have classified you (for example "penny-pinching empty nesters" [parents whose children have left home] or "fanatical video-game players"), classifications of your value as a customer ("most valuable" or "below-zero customer"), and data they have rented or acquired from other companies ("creditworthiness = questionable").

The need to protect competitive secrets

As Varney pointed out in the Senate hearing, companies can spend a great deal of money acquiring this sort of knowledge, and they are fearful of giving up the competitive advantages they hope to gain as a result. For example, by disclosing how they classify individuals, they are revealing to competitors the same information; what's to stop dozens of your employees from mystery-shopping a competitor's site and then requesting disclosure of their profiles? Once you know how a competitor differentiates among its customers, you know how hard you need to work to leapfrog them.

Increasingly, information will be the most valuable asset a company owns. In the normal course of business, firms learn a great deal about which offers worked, which suppliers are most effective, how best to motivate employees, how to streamline business processes, what the effect of seasonality has on purchase patterns or delivery reliability, how raw-material prices fluctuate, and when to minimize – or pump up – market expectations. To think about it another way, this information represents the memory of a corporation's life. Some executives believe it is absurd to suggest that companies can't remember or use the information they learn in the normal course of business.

Dateline: 2014.

Internet News Alert: COURT ORDERS MEMORIES TO BE REMOVED FROM SCIENTIST'S BRAIN?

In what is certain to be a landmark case, Dr Alyssa Paige, a scientist in London, sued her former partner for possessing and using without authority confidential information and asked that the court order a medical procedure to remove memories from her partner's brain.

Such a removal process is technically possible, because both the scientist and her partner had supplemental memory chips implanted in their brains and the disputed information – which largely comprises complex mathematical formulas – is stored in these chips. Dr Paige contends that the memories are personal information which she has a right to control under the Global Act to Protect Individual Data.

"I shared this information with Dr Skinner when he was a loyal friend and colleague. We had an implied agreement that he would never use such information to harm me, but his recent actions indicate he is doing exactly that, so he no longer has my permission to store or use my information, either on his computers or in his memories."

There is no clear line between what organizations have a right to remember and what people have a right to protect and control. Most would agree that the above example is absurd – no one has the right to order that someone "forget" information in their memories – but that is exactly what many propose that corporations do. That prospect is hugely troubling to many executives.

Inability to comply

To my ears, the merchant cited in the Senate hearing is Amazon.com, which arguably is one of the most advanced e-commerce competitors in the world. It's a mistake to assume that others can match Amazon's capabilities. Even if a government passed laws saying the company has to reveal all the data it holds on a given individual, most companies could not comply, except at a ridiculous cost. The vast majority of firms have yet to link all their databases or even create a common approach to storing customer information. Many of my clients have been unable to determine whether the "James Smith" contained in the books database is the same person as the two "James Smiths" in the electronics database or the "Jim Smith" in the customer support database or the "James Smith Jr" on the free email updates file. In fact, one of the key forces driving the CRM (customer relationship management) movement is the need of firms to clean up and link their databases.

Privacy advocates have differing motivations; they range from extreme views that advocate complete anonymity to more balanced efforts to protect mainstream interests. To many people, Jason Catlett's point of view, which seems to resonate with Senator McCain, seems completely logical. Why can't people have access to the corporate profile that companies accumulate about them?

Many voters will dismiss as gobbledygook attempts to explain why this doesn't make sense, and perhaps they should. You can make a strong case that, if companies really need to store personal information, then they have a duty to work just as hard to give people access to this information.

If this debate gets emotional, and it will if enough major privacy scandals break, the popular appeal of this message may overwhelm all others. That will be unfortunate, because it is hard to create reasonable policies around technologies that most people don't understand, including some executives of the companies implementing the technology.

One thing is clear. Technology will change faster than the laws. The pace of this change makes it extremely difficult to avoid creating legislation that will inadvertently forbid practices no one intends to prohibit. Jacob Palme, a professor at Stockholm University, has outlined worthwhile concerns about Swedish law based on the European Union Data Directive, which basically comes down to serious infringements on the freedom of speech "supposedly" protected by the Swedish constitution as a result of new privacy laws. He contended that the law "is so generally worded, that it actually applies to almost all human activity, since almost all human activity to some extent involves personal data in computers or on paper."

Palme gave use of search engines as an example of an activity that is illegal according to the act.[7] For example, if it is illegal to use my personal data without my permission, then every search engine that provides a list of information about me is technically violating such an act. I just looked up my name at one search engine and received 2,008 citations back, none of which I authorized. Of course, in practical terms, no one authorizes search engines, but this illustrates the challenges of balancing a person's right to privacy with society's constant search for knowledge and insight.

What If Opt-in Becomes the Law?

As of mid-2001, in most countries and most business situations, opt-out is the generally accepted business practice. "Opt-out" means that a person must *object* to a use of personal information in order to stop it. It's a quick way of saying, "We're going to do this unless you object." A simple example of opt-out is requiring people who register at a company's website to find and check a little box if they *don't* want to receive product updates via email. At the present time, many laws agree that sufficient protection exists for individuals as long as companies comply promptly with a request from people to terminate the use of their personal information. Opt-out places the burden of work on the individual, not the company. As data trails proliferate, the volume of this work increases dramatically.

It's getting harder for individuals to opt out; it's hard to keep track of all the companies that are trying to track you. I recently changed the security settings on my Web browser so that no cookies could be placed on my hard drive without my permission. Then, every time a website attempted to place a cookie, a warning box popped up and asked my permission. On some sites, I was forced to hit the "no" button five times on a single page, and repeatedly each time I loaded a new page. In the majority of cases, these requests preceded any indication on my part that I wished the site to remember information about me; I was just browsing.

My first reaction was annoyance at the number of attempts to track my activities that previously happened without my permission. But it didn't rise beyond annoyance, because with the current technology I can instruct my browser to reject all cookies. If I were really upset, there are also a variety of software applications that would enable me to change my settings for different sites, depending on whether they are trusted favourites or not. My second thought was how inconvenient it can be to be constantly asked for permission. On some sites, I was constantly hitting the "no" button to ward off multiple attempts to place cookies on my hard drive. At other sites that I trust, I had to keep hitting the "yes" button. I've since switched my browser settings back to "accept all cookies" and periodically open my cookies folder and delete those that I don't wish to keep on my computer.

For most people, it's unclear – even to them – which they value most highly: convenience or privacy. In theory, we want privacy, but what most of us really want is insurance against harmful abuse. You don't want to be discriminated against, threatened, or fired because of how firms manage data about you. Most situations are far less concerning. The challenge is creating laws that provide protection without creating greater problems than those the laws were designed to prevent.

The opposite of opt-out is opt-in. "Opt-in" means that companies will not do anything until you give your permission. Some believe that opt-in should become the new standard worldwide. Canada has already passed the Personal Information Protection and Electronic Documents Act, which will be implemented in three stages; the first, affecting federal works, started on 1 January 2001. The third stage, to commence 1 January 2004, is the most impactful. It "extends to the collection, use, or disclosure of personal information in the course of any commercial activity within a province."[8] It requires that "organizations covered by the Act must obtain an individual's consent when they collect, use or disclose the individual's personal information."

This law requires that individuals opt in and forbids the opt-out practices that most firms use today, especially online. It even goes so far as to say that if information is used for a purpose other than the one to which a person consented, the organization must obtain consent again. Since we are still years away from the law's full implementation, it's unclear how significantly it will impact Canadian businesses. There's a big difference between writing a law and enforcing it. What's clear, however, is that if such a law was implemented and enforced, it would change the nature of business relationships.

Opt-in Relationships

In an opt-in world, individuals need an incentive to act. Think about all the catalogues you receive in a year and imagine that you enjoy every one of them. Now imagine that the laws change and that every one of these catalogue merchants is forbidden from mailing you a catalogue unless you specifically instruct them to do so. How many merchants will you forget to contact? Even if such merchants send you handy cards or email that only requires you to check a "yes" box, I'll bet you'll miss many of them.

Now, remove my generous precondition, which is that you enjoyed every catalogue you received last year. In my house, we spend half our time sorting through the mail and shaking our heads at the junk we receive. There's no way we are going to request a catalogue from the How-To-Profit-From-The-Coming-Technology-Crash/Boom folks, or the 1001 Ways to Use Baking Soda conglomerate.

The simple process of thinking about which companies you give permission to contact you will force most people to edit their merchant circles. Do you really need catalogues from Dell, Compaq, Gateway, and PC Warehouse? How many fruit-of-the-month club solicitations do you want? How about credit-card offers?

Even within a firm or a distribution channel, it's likely that people would use a migration to an opt-in world to limit the barrage of information they receive each day, eliminating updates from other departments and business units.

Without significant additional incentives, most people would refuse to permit many of the practices in which companies engage today. See how many of these practices you would approve, if given the choice by one of these four imaginary companies:

1. Acme's Secrets would like to rent information about your underwear purchases to 46 merchants in the next six months.
2. You run a group of retail stores, and Acme Manufacturing wants to publish some of the sales figures collected from your stores in a Sales Tips packet it is distributing to other retailers, including some with whom you compete.
3. Your employer, Acme Conglomerate, wants to compare the number of days you were sick or on vacation to its other 4,000 employees to see how your job-performance ratings measure up and determine whether people who use all their vacation days perform at a lower level.
4. Acme Vacations wants to show you promotions for trips to the Indian sub-continent – instead of to France or Switzerland because it noticed that you live and work in neighbourhoods that have high concentrations of Indian nationals.

In all four of the above examples, there is no benefit for the person, only for the company and other of its stakeholders. This is fairly typical of the way most firms do business today. They gather data and use it *en masse* to benefit their own company. That's why most companies object to opt-in policies, because they know few people would grant permission, if asked, or that to obtain such permission they would need to "sweeten the pot". But in an opt-in world, services can't just be sugarcoated. They would have to be fundamentally reshaped to be in both parties' interests. This takes a different way of thinking, one that is foreign to most executives, who have never before been held accountable for the interests of the people they serve, but largely only for the business results, such as higher sales or lower costs.

Here are four more examples, but each of these describes a practice that benefits *both* the imaginary company and the person involved:

1. Happy Secrets wants to remember the items you really want, but that are too expensive for you at full price. To do this, they'll also need to know your size and colour preferences. They will send you an email if they are ever able to sell one or more of those items at the price you can afford, for example, the size eight grey cashmere V-neck for £80, instead of £150.

2. You run a retail store, and Happy Manufacturing wants to remember the products you carry and the prices you pay (including your special volume discounts), so that it can eliminate promotions and literature about products you don't carry from its communications with you, saving your team time and money.

3. Your employer, Happy Conglomerate, wants to remember the names of the clients with whom you exchange email, so that it can produce "maps" of contacts between its employees and those of its clients, making coordination easier for everyone and identifying best practices of account teams that can be adopted by other teams.

4. Happy Vacations wants to remember which trips and accommodations – you liked and disliked, as well as where you aspire to travel in the

future, so that it can recommend best deals that you are most likely to appreciate.

In each of the cases above, the company succeeds, and probably makes a greater profit than it did under the original approach. For example, Happy Secrets will be able to liquidate excess inventory without ever having a public "sale", and it may even be able to order inventory more aggressively, because it will know approximately how many items it can liquidate simply by honouring discount requests. Happy Manufacturing is making it easier for distributors to do business with it than with competitors, which should increase sales. Also, by listening to distributors' feedback concerning items in which they don't have interest, the manufacturer is increasing the odds that distributors will listen more closely to promotions about the items they do carry.

If I could only convey one message to the CEO of your company, it would be this: you must follow the lead of examples such as these, and find new ways to satisfy the needs of individuals while you simultaneously meet the needs of your own organization. This is the only way to balance profits with privacy, and to minimize the need for further legislation, which will be certain to create a more restrictive and more difficult climate for business.

There is fertile ground where information and innovation meet, and the most successful companies will recognize that new sources of information must be used to create win–win opportunities for both a company and its stakeholders.

In the long run, opt-in laws could force businesses to adopt much more logical business practices, but the transition could be a nightmare for both businesses and individuals. Companies do nearly everything without permission today, and switching gears without slowing down the global economy would be extraordinarily difficult. Here are some of the common practices and services that could be prevented or slowed by opt-in legislation:

- publishing phone directories;
- publishing membership lists of community organizations, which are commonly used to make it easier for members to contact each other;
- posting test results on school bulletin boards;
- sending product upgrade notices to people who purchased a now-outdated product;
- distributing complete race results following a major marathon;
- sending free magazine or credit-card offers via snail mail;
- inviting executives to attend new business conferences;
- personalizing Web banner advertisements;
- publishing retailer locator lists;
- cross-selling product lines; for example; promoting printers to customers who purchased a personal computer;
- distributing student comments on a professor's performance;
- Internet search engines.

There are obvious problems here. The first is that privacy protection is difficult to enforce when one party has leverage over the other. As a precondition of employment, companies could easily insist that employees sign an agreement that allows the firm to publish information about them. A private school could do the same with both its teachers and students. It would be easy for large companies to force suppliers to approve most uses of data, because the alternative would be losing the large customer.

The second problem is that a poorly written law could prohibit practices that few intended to forbid, such as search engines and telephone directories. Most would agree that privacy protection shouldn't infringe on free speech.

A third problem is paperwork. No one seeks to make companies less efficient or to burden them with excessive bureaucracy. But we could stumble into a situation in which companies spend so much time obtaining and managing permissions that their basic business becomes unprofitable. UK businesses are already struggling under the weight of existing regulations; small and medium-sized firms are already at the limits of their capabilities. Today, most businesses have no way of attaching a person's "handling preferences" to particular pieces of data about them. The preferences they do collect tend to be yes or no: If you don't want to receive mail from us, we'll take your name off the mailing list. If you don't want this additional newsletter, we won't add your name to that list. But few companies could accommodate a request such as, "You can share my grocery purchases with food manufacturers but not with insurance companies or prospective employers."

Inside-out/Outside-in

Companies need a simple solution to balancing privacy with personalization. Most of the proposed approaches to protecting privacy suffer from one of two significant flaws. They are too feeble, offering no significant protection to individuals, or they are too complicated for either individuals or companies to understand, never mind trust.

Here's the privacy policy for HowPersonal.com, the Web portal I founded to educate customers and employees about privacy and personalization:

We want to remember you, but never intrude. So any information we remember about you is used within our organization to make your next interaction with us more convenient, unless you instruct us otherwise. We *never* share personally identifiable information about you outside of our company unless you consent in advance.

This policy is 51 words, which is substantially shorter than the 4,910-word disclosure agreement Citibank customers must agree to before using its online banking site. To understand Citibank's, you need (a) a lawyer, (b) an assistant, or (c) much too much free time. Many companies are allowing their privacy policies to grow to this absurd size. It protects the company, but sends a message to everyone else that the company is playing the role of a crafty lawyer, rather than a trusted collaborator. In reality, most people click "I agree" without even reading the text. They discover flaws only when problems arise later.

Our policy is based on a two-pronged standard we call Inside-out/Outside-in. It attempts to balance the needs of a company with the best interests of an individual. For information used within the company, we have an opt-out standard. That is, we assume it's OK with you if we remember information about you. But if we want to release information about you to outside companies, we use the stronger opt-in standard. This is because release of such information can cause you unintended problems or concerns, and you have a right to approve such actions in advance.

Inside-out/Outside-in Privacy Policy

Outside firm:
opt-in
(*or* make data
anonymous)

Inside firm:
opt-out

"Outside" firms means allowing other firms to use personal data for their own purposes. It does not mean retaining service providers to support the company's own activities. For example, companies regularly employ other vendors to send emails or snail mails to its customers, but these companies do not get to use any personal data for their own purposes. In such situations, companies could apply the weaker opt-out standard.

There's one other element to this policy, which is that we can use and share information about the individuals with whom we deal, as long as we first remove any personally identifiable information. Although this sounds like a sly loophole, it's necessary to preserve the interests of not just the company, but also of the larger community a company serves. Here's an example:

A 66-year-old woman gets severe stomach and chest pains while on a camping trip. Her portable medical alert sends messages to her doctor and the closest paramedics, who locate her in the park using the global positioning system on her wrist transmitter. Once in a local emergency room, she gives the doctor permission to access her online medical record.

Comparing past and current medical tests, the doctor rules out heart trouble and diagnoses viral gastroenteritis. Soon after, the local health department's automated surveillance system, which collects anonymous data from providers, recognizes a cluster of tourists with similar symptoms. Officials trace the problem to a punctured sewer line, which has contaminated park drinking water.

This story isn't possible – yet. It was invented by the National Committee on Vital and Health Statistics, or NCVHS, an advisory group convened by the US Department of Health and Human Services. The idea was to demonstrate how a so-called National Health Information Infrastructure would work. Such a network would be used by patients, doctors, researchers, and public-health agencies to share electronic medical records, monitor far-flung patients, and participate in early-warning systems for public health dangers. It could also help reduce medical errors and improve diagnostic accuracy.[9]

Anonymity balances a company's right to leverage its information assets with an individual's right to privacy. In the vast majority of cases, anonymity protects an individual from invasions of privacy. That's what makes it possible for doctors to diagnose patients faster, without invading the privacy of any single patient. The exception is where others can pierce anonymity because the person's identity becomes obvious, thanks to deductive reasoning or a small volume of data variables. In the spring of 2001, Continental Airlines was embroiled in a dispute with some of its biggest corporate customers, because the airline insisted that such customers disclose how much they are spending, ticket by ticket, with other airlines. A few large companies, such as American Express and Johnson & Johnson, dropped Continental as a preferred provider due to this requirement. Johnson & Johnson claimed that its agreements with other airlines forbid it from making such disclosures, but Continental argued it needed such information to verify that customers were complying with agreements that, for example, offer 25% discounts to customers that used Continental at least 60% of the time.

Continental argued that customers can comply with the restrictions of other airlines by masking the airline names before sharing the required data. But Marianne McInerney, executive director of the National Business Travel Association, said, "The truth is, you can't mask this data. It's very easy to see who the other airlines are and what their rules are." This is because many routes are flown by just a few airlines.[10]

The airline also argued that the data it sought was the same information it had long received through travel-agent reports and that it was only seeking more specific data to be filed more quickly. If true, this illustrates one of the dilemmas of companies as they confront new technologies: Others challenge the use of new technologies to automate practices that have long been commonly accepted. This is because many such practices operated under the radar of formal policies, as a result of informal activities, or in unglamorous areas of a firm that attracted no attention, especially from the press. Today, the press and privacy advocates consider interactive and database technologies to be highly charged and potentially intrusive. Many watch carefully for hints of scandal.

At the time of mid-2001, privacy laws were evolving rapidly and starting to overlap. In today's global economy, no one nation or group of nations can legislate in isolation. The European Data Directive, for example, which is more restrictive than laws in some other nations, forbids member companies from transferring personal information to companies in nations that lack sufficient safeguards to protect such information. It is impossible to predict how laws will evolve, although it seems certain that laws will move in the direction of increased safeguards, which means additional complications for businesses.

In this environment, Inside-out/Outside-in offers a prudent and practical course. For most companies, adopting this approach will mean higher incidents of opt-in practices, which will strengthen relationships and teach the firm to build win–win relationships. If individuals decide not to opt in, a firm retains the right to use the information after eliminating any personally identifiable details. For internal purposes, most companies can continue to do business as they have in the past.

Why You Can't Wait for the Laws to Change

In Chapter 1, we looked at Don Peppers' and Martha Rogers' concept of a Learning Relationship, in which a company learns so much about how an individual prefers to be treated that the company makes loyalty more convenient for the person than disloyalty. Once a firm has created such a relationship with its most valuable stakeholders, these people will prefer to work with that firm, rather than with a competitor, because working with a competitor would mean reteaching another firm how to accommodate the qualities that make them unique. Unless a firm experiences a disaster – it misses a technological discontinuity, allows product quality to slip dramatically, or suffers from fiscal mismanagement – Learning Relationships offer it a potent competitive advantage.

Firms that adopt the Inside-out/Outside-in framework, and that use personalization to develop Learning Relationships, will be preparing for strengthened individual protection while they simultaneously lock in stakeholder loyalty. Once the laws change, other firms will have to adopt similar practices, but these practices will *not* result in similar loyalty. Here's why.

If a person already enjoys a Learning Relationship with an employer or merchant, the person has no reason to switch to a company that suddenly develops the exact same capabilities. It doesn't matter that your firm is capable of supporting a Learning Relationship; it only matters that your firm is the first to do so with a valuable stakeholder. There's no effective strategy for leapfrogging an entrenched competitor protected by such a relationship, other than hoping the competitor encounters a massive disaster.

Thought Exercise . . .

Opportunity or Threat?

None of us can predict changes in the privacy laws, but you still need to anticipate potential changes so as to better understand how you can, and can't, use technology to strengthen business relationships. Here are five scenarios, all plausible and significant in their potential impact. For each of them, consider what you could do to prevent or react to the scenario. Or just read them quickly, but remember that the future will be much messier than any of us anticipate. This is just a taste of changes to come . . .

Scenario 1: The countries in which you operate pass laws that require a person to opt in before you can store personal information about them, other than their name and address or information that you are required to collect by law. Starting two years from today, you must comply with this law, which means you must obtain written permission by then to continue storing information you have already collected and to be able to store information you wish to collect once the law has been implemented.

Scenario 2: A human being is cloned without advance knowledge or permission, and the procedure goes horribly wrong, resulting in a global rush to protect personal information such as DNA, fingerprints, iris patterns, and even facial characteristics converted into data files. The public not only refuses to use such identifiers, but also backs away from firms that were already using such methods.

Scenario 3: A FTSE 100 company admits sharing confidential data from its leading suppliers with other suppliers, without their permission. This practice was going on for 18 months, and four suppliers allege they were forced into bankruptcy by this practice. As a result, suppliers begin re-examining with renewed intensity data-sharing practices with their largest customers.

Scenario 4: A semi-organized group of hackers breaks into the human resource databases of eight leading firms and acquires employee names, mobile phone numbers, credit card numbers, address book contacts information, and health insurance claims data. The hackers begin to circulate random and recurring emails that distribute this stolen information worldwide. Two days later, they publicize a list of additional target companies, and yours is on it.

Scenario 5: Your firm releases data to a third-party data provider, after eliminating personally identifiable data. But a disgruntled former employee, who knows well the flaws in your system, posts simple five-step instructions on Web discussion boards that will let any company or person identify 20% of the people described in the data you released. (Your identification system used home phone numbers to identify individual accounts. The first three numbers of each phone number were deleted before the data was released, but not the town names. In any small town that has just one exchange, people can replace the deleted digits, then use any Web-based reverse look-up directory to get names and addresses.)

Notes

1 http://junkbusters.com/ht/en/testimony.html
   I sent copies of the excerpted testimony to all the participants and offered them the chance to comment on this interchange, correct any inaccuracies, or to object to its publication. None objected.

2 Press release
   http://viisage.com/january_29_2001.htm

3 "Winter Olympics Group is Considering Super Bowl's Controversial Surveillance," *Wall Street Journal*, 5 February 2001. Subsequent to this article, the Olympics Committee has reportedly decided not to use this surveillance technology.

4 "Police Video Cameras Taped Football Fans: Super Bowl Surveillance Stirs Debate," by Peter Slevin, *Washington Post*, 1 February 2001.

5 Written by "Onedingaling", 10:26 a.m., 1 February 2001
   http://forums.prospero.com/n/main.asp?webtag=wpnation&nav=
   messages&msg=1464.1

6 Redryder001
   http://forums.prospero.com/n/main.asp?webtag=wpnation&nav=
   messages&msg=1464.1

7 "Concerns Regarding the EU Data Directive," distributed to **Politech@politechbot.com** mailing lists, 30 November 2000.

8 "A Guide for Business and Organizations from the Office of the Privacy Commissioner of Canada: Your Privacy Responsibilities," Canada's Personal Information Protections and Electronics Documents Act.

9 "Health-Policy Groups Put Data Network on Agenda," by Laura Landro, *Wall Street Journal*, 9 February 2001.

10 "Continental Airlines Loses Some Accounts After Data Disclosure Demand," *Wall Street Journal*, 6 February 2001.

4

Whose Agenda?

In the 1987 movie *Rain Man*, Dustin Hoffman gives a convincing performance as Raymond Babbitt, an autistic savant who has fantastic mental powers, but no real consciousness. Many of the companies working to personalize the way they interact with people make me think of Raymond. They are very smart, but can't maintain anything resembling a relationship with a single human being. This is because, as Don and Martha point out in the foreword, most companies begin with the question of how technology can be used to extract more money from a person, instead of asking how it can be used to better serve the needs of people vital to the corporation's long-term success.

Want to be the next Richard Branson, or make your already-established company the next GE? Sync your company's activities with the needs of successful individuals, so that each person views your company as watching out for his or her interests. By picking the right people at the right time, your company will profit hugely. This is the potential of making it personal.

So far, most personalization initiatives have focused more on corporate, rather than individual, needs. That's why we are seeing heightened sensitivity to privacy concerns, and also why many personalization initiatives fall short of their goals. As we look at the technologies that make personalization feasible, through the eyes of the pioneers inventing these tools and bringing them to market, you'll gain a sense of what it takes to balance the needs of individuals with those of a company.

The Early Days of Personalization

To generalize, personalization involves one or more systems that gather, organize, filter, and/or deliver information. Information delivery can be the end goal in itself – such as in the case of a news site – or it can be the trigger for other actions, such as the creation of a customized product that is then delivered to a customer. In most situations, more information improves an organization's ability to meet individual needs, if the organization creates the right processes.

Since technology has tremendously expanded the ability to gather and organize data, it's now possible for large companies to deliver more relevant services than any small firm can. This doesn't mean more friendly, or more personal in the human sense ... even at their best, such firms can't match the experience of working with an entrepreneur you've known for 20 years. It means more useful and more tailored service.

Say you are a 15-year-old school student living on the south coast of England and you are fascinated by Chinese art of the 15th century. Odds are you have no one that you can talk to, who can give you advice or can tell you something you don't know about Chinese art. But then you go to a large college and discover there are actually courses in your favourite subject, and your classmates and professors expose you to other types of art that are equally fascinating but that you never knew existed.

This is what happens when your neighbourhood gets larger. It can accommodate a wider range of interests and still find people with needs in common. Firms that leverage the knowledge inherent in their neighbour-hoods can accommodate, at the finest levels, the needs of individuals. Geography is only one way to define a neighbourhood; you can now build neighbourhoods around any problem, opportunity, or shared interest.

Said another way, when opinions matter, more opinions are better. Tim Zagat, founder of the famous Zagat restaurant guides, sums this up by asking, "Wouldn't you rather have 3,000 opinions on whether a place is good than just one?" Opinions matter when you are operating in unfamiliar territory, such as launching a new project or flying into a city you don't know well. They matter when you are making an important decision with long-lasting financial or personal consequences, or when you need to be especially precise in your actions, such as when configuring a network to connect your European offices.

The challenge with opinions, of course, is knowing which ones to trust. Among your network of friends and family, you develop a sense of who knows what. My wife, for example, is a wonderful judge of character but a lousy critic of automobiles. When I take on a remodelling project, there are three or four people who provide me with consistently invaluable counsel. I also benefit from anti-guidance, which is knowing that if certain friends love a movie, I'll hate it, and vice versa. Epinions is a recommendations site that attempts to solve this problem by letting readers rate the opinions that people provide. You can go to Epinions.com and write an opinion on just about anything: cars, TVs, books, software programs, or skis, etc. Then, when you need someone else's opinion, you search for that item and get a list of all the available opinions. You also get a detail that shows how many others trust this reviewer, and you can see how many reviews he or she has written.

A prudent person is sceptical about recommendations. As Ronald Reagan said, "Trust but verify." Many merchants accept payments from manufacturers to promote certain items or to make recommendations of these products more visible than others. The best opinions come from sources that you both trust and know to be expert. One of the best ways to determine trust is to examine whether a person has taken the action he recommends to others. There are a variety of technologies that facilitate this sort of opinion sharing, and it will be useful for you to get a sense of them.

Net Perceptions[1] is a software company that was an early pioneer in the recommendations field, and it uses a technology known as collaborative filtering, which works like this in a merchandising environment: When you come to a website or call a company, the software compares all of your purchases to date against the company's database of other customers' purchases. The software looks for patterns, trying to find other people who have purchased the same products you have. Then, it looks at these similar people and finds products they have bought that you have not yet purchased. It recommends these items to you.

All of this happens in a split second, and the recommendations are specific to the exact moment you visit the site. Come back in a month and the odds are some of the recommendations will change, because new products come out, some of which are being purchased by people like you.

Remember Senator McCain's logical question: Shouldn't a person have access to the profiles companies keep on them? With collaborative filtering, which frequently is the technology that enables book and other retailing recommendations, there are no static profiles. A retailer's list of the books you are likely to buy is displayed to you as a list of recommendations. There is no other "profile" lurking behind the scenes that drives this process, although the company does store other information about you, such as the products you have purchased, the credit cards you have used, and so on. Generally, you can access this information, too, via a button such as "Your Account". In the senator's case, there is no note in his file that says he likes history books. There is not a long list of instructions that says, "Offer him these three books." The recommendations are the result of a process that happens in real time.

That having been said, Net Perceptions software includes a number of features that enable its clients to have significant control over the way that the technology serves their customers. The loftily named Serendipity Control function enables companies to reduce the frequency that the software recommends especially popular items. If 45% of customers rent whatever movie comes out this week, a firm doesn't add any value by including that title in a personal recommendation. Some companies use Serendipity Control in the opposite direction, to increase the frequency of obscure items if they think that there's a reason for them to be brought to the customers' attention. Other capabilities are designed simply to deliver quality service to a customer. Synonym Rejections says don't recommend large eggs *and* jumbo eggs to the same customer. Either one is enough.

Net Perceptions tried to apply its technology to the practice of knowledge management, focusing on certain industries, such as financial services and pharmaceuticals, where firms manage vast amounts of information and face significant challenges to eliminate redundancy and encourage reuse. Sang Kim headed up this business for Net Perceptions. Despite his firm's highly visible efforts to promote personalization, Sang said, "I don't think about personalization. I think more of finding stuff that is hidden and figuring out a way to get it to people."

Just as opinions used to be much harder to obtain in the quantities that they now flood the Web and intranets, so, too, has most data in the world been hidden until now. Public records, for example, are just that and have long been theoretically available to, well, the public. But they have generally existed on dusty, often mouldy pieces of paper stuffed into ancient filing cabinets crammed into offices that are painted, as Tom Wolfe would say, "good-enough-for-government green". Highly motivated people could get the data, but most of us are either not highly motivated, or too busy to access the data. But now public records are being stored on computers that are being networked together and attached to powerful search and analysis tools like the ones Sang Kim marketed.

A National Public Radio commentator, Katharine Mieszkowski, pointed out, "Most of the juiciest details of our friends', neighbours', and enemies' lives are a matter of public record, and that information is finding its way online, fast ... Court records, filings for bankruptcy, home-sale prices, marriage and divorce records, voter-registration information, criminal records, are all making their way onto the Web." The Internet Open Records Project (*www.openrecords.org*) believes that access to public information is essential to a free and open society, and thus its site today enables people in Dallas, Texas, to access online detailed records of who voted – and who didn't – in each election. Their goal, of course, is to enable this capability everywhere. Only the nosiest person would ever have spent days at the Public Records Office investigating whether a neighbour voted consistently, but now anyone can come home from a frustrating public meeting and learn in 30 seconds that the loudmouth in the front row never bothered to vote in the three referendums to which he now objects.[2]

But for all the information already contained in databases, Sang Kim said that the most important information still exists in people's heads. "How does a search engine eliminate redundancy? How will it say, 'Stop! Other people have done this.'? It won't. It just parses your question and retrieves information." The missing factor, said Sang, are opinions. He said, "When you reuse stuff, you want to reuse stuff that's valuable, not just all stuff. Value is a human judgement. You only get this through feedback."

Net Perceptions sought to automate a process that could be done manually. If a company could shut down every night, when everyone would then talk to everyone else, it could start up the next morning with perfect knowledge about what worked and what didn't. You wouldn't need software. Since this isn't practical, firms such as Net Perceptions are racing to motivate people to share what they know, so that this information can be used to help people with similar challenges and opportunities.

Tacit Knowledge Systems is also focused on this problem, but is going at it from a different perspective: email. Tacit's software sorts through email messages and corporate documents to determine who knows what. The firm's name refers to the useful stuff each of us knows that no one bothers to ask and is hard to access normally. Tacit is not directly searching for opinions, but instead is trying to identify people with specific expertise.

Unless an employee uses an opt-out button to exclude an email from review, Tacit's system sorts through every email looking for content that will reveal what the sender knows. It weighs information differently, in an effort to ascertain the level of knowledge a person has. For example, if every other email a person sends includes the words "statistical" and "statistics" then it is a pretty good bet that person knows a great deal about statistics. If, on the other hand, someone occasionally mentions the words, and usually only with the word "football", then that person shouldn't be identified as a statistics expert.

In an effort to allay privacy concerns, Tacit software doesn't reveal, or even store, the content of emails. Instead, it uses the information to build a private profile of each employee, which only the employee him or herself can see. But when other employees need help on a particular subject, Tacit's system will identify relevant experts and check to see if they are willing to step forward voluntarily. The notion is that, without the profiles, it would be impossible for employees to find many of these experts. Not all profiles are private. Each employee can also maintain a public profile, which makes it easier for others to locate them.

No company installs such software out of an altruistic desire to help people. The vast majority of personalization initiatives to date have been designed to increase the effectiveness of marketing and merchandising or, to a lesser extent, to better share knowledge within a company. So most implementations aren't as valuable to individuals as they could be.

Christian Gheorghe, founder and CEO of Tian Software, is dismissive of many personalization approaches. "Today, the model of personalized ad serving is the equivalent of junk mail," he said. "There's a pool of let's say 12 banners, which someone created by guessing what customers like, and they throw them out and see what sticks." He argued that collaborative filtering and other grouping mechanisms are mainly being used to cross-sell and upsell customers. Instead, Gheorghe said, firms need to be able to stage experiences that are relevant for a certain person at a certain point in time. You can't do this by predicting in advance what a person will need, because no person or system is capable of making such predictions accurately, especially when dealing with thousands or millions of people.

Tian's approach is to allow his personalization system to make decisions in real time, as a person is interacting with it. In other words, instead of telling a computer "offer skis to everyone who buys ski boots," Tian's system might be programmed "once a person has made a purchase, consider the item they just purchased and their other actions so far and then decide what to do next." This approach, said Gheorghe, "Forces rules to live or die based on whether they work." By rules, he means instructions that tell the system what to do. Both of my examples are rules, and the challenge is determining which of the rules produces the results a company seeks. The individual's needs are secondary to this larger corporate goal. No one in a big company says, "Yes we missed our sales targets again, but on a far more important note, we made Jimmy Doherty really happy."

Lisa Hammitt, CEO and CTO of Black Pearl, another personalization technology firm, questions whether personalization should be a goal unto itself, or rather a by-product of extending the way companies already do business. Here's her pitch to leading financial institutions: "You have data and know-how trapped in the walls of your enterprise. It isn't getting out there. We want to help you extend your value-added sales tactics to the virtual world. We want to enable that. You already have a successful business model that made you a leader in your industry, and we can help you extend those same best practices into the electronic channel. To do that you need a collaborative approach to customers, driven by courtesy, and powered by a timely flow of information to customers. It means putting lots of technology in place. You have data about customers in many places. Logic about how you deal with customers is elsewhere. Increasingly, you have multiple touchpoints, including mobile, Web, phone, face to face, and mail. Our role is to turn data into something much more useful to someone doing business with you."

When you translate this pitch into action, it often means taking the business logic that already drives financial advisers – such as recommending conservative investments to people over the age of 60 or aggressive stocks to young couples with plenty of time to be patient – and building it into systems that can help brokers make better recommendations to their customers and can speed a company's ability to respond to an individual's need. This isn't a revolutionary approach, but it can serve to reinforce an existing relationship without threatening either the business or the people they serve.

Embracing the Individual's Agenda

Tony Durkins was late again. Unfortunately, today was Valentine's Day and he was headed to dinner with his wife with no flowers and no gift. His one hope was his phone.

Turning the corner out of his office parking lot, he pressed two buttons on his phone, accessing the personal help menu. "Do you need a Valentine's Day present?" it asked. Recognizing his tendency to procrastinate, he had programmed the system to guess his intent on special occasions.

"Yes," he answered.

"Flowers? it prompted, suggesting the highest probability solution first.

"Yes," he responded.

The system posted two florists on the phone's display screen, along with the distances to each from his current location. The first one was labelled "closest" and the second "best bet," which indicated that it offered the best combination of proximity, quality, price, and selection, based on Tony's past buying preferences.

He was too late to go for the best bet, which was an additional two miles away. He chose number one.

"Do you need directions?" asked the system. Tony responded by switching off the phone, so he wouldn't get diverted by more client calls. He was just two blocks away from the closest merchant.

The Web enabled our current brand of corporate-driven personalization, but the proliferation of mobile communications devices will bring a new wave of personalization that is more obsessed with the needs of each person. Ads on a television or computer are one thing, but when a person's mobile-phone screen is two inches square, his or her tolerance for advertising goes way down. Mobile devices lack the ease of interacting or the bandwidth to allow much real time exchange of data. Only a monk has the patience to type in name, address, credit-card number, and order using a telephone or even a Palm Pilot keypad. To get around this problem, people need some way to store their personal information and transmit it with a single command. There are two candidates to store such information: a trusted agent, or the user himself. Let's look at the personal option first, to understand its benefits and limitations.

PERSONAL PROFILE
LEVEL 1 INFORMATION (PUBLIC – OPEN TO ALL)
NAME: Matthew Richard

HOME LOCATION: SW1 9SP

WORK LOCATION: E17 9QH

AGE: 29

MARITAL STATUS: Single

CHILDREN: None

OCCUPATION: Java programmer

DRIVES: 2001 Porsche Boxster

ACCEPTS UNSOLICITED EMAIL: No

WILLING TO RELEASE INFORMATION FOR COMPENSATION: No

LEVEL 2 INFORMATION (PRIVATE – FOR PURCHASES ONLY)

CREDIT CARD NUMBER: XXXX-XXXX-XXXX-XXXX

CC EXPIRATION: XX/XX

CC TYPE: Visa

SHIPPING: 21 Ashdown Road, Fulham, London SW1

BILLING: Same

PHONE: 0207 431 286

DELIVERY REQUIREMENTS (default): Normal ground delivery

DELIVERY INSTRUCTIONS: Leave packages with no. 23

MAILING LIST PREFERENCE: No mail

LEVEL 3 INFORMATION (PRIVATE – ACCESS BY PERMISSION ONLY)

LIKES (KEYWORDS): Karate, windsurfing, Japanese joinery, Neil Stephenson, AOL, bicycling, chess, Formula One, horror movies

DISLIKES (KEYWORDS): Kids, golf, politics, globalization, advertising, processed foods, cats, DIY, alcohol, smoking

WEB FAVOURITES: OK to share with trusted friends

RECENT PURCHASES: OK to share merchant names with trusted friends but not items purchased

LEVEL 4 INFORMATION (PRIVATE {BUSINESS} – ACCESS BY PERMISSION ONLY)

EMPLOYER: Pixel Studios

UNIT: *Motion capture*
PROGRAMMING LANGUAGES: *Java, C++*
WORKS ON: *Apple Titanium PowerBook G4*
RELATED SKILLS: *Drawing, colourizing, voiceovers*
REPORTS TO: *Robin Freedman*
SUBORDINATES: *None*

No individual controls the processing power necessary to gather and filter large volumes of data, but we all have sufficient power to store our own information and release it as we see fit. If Matthew has a software program that allows him to set up a profile such as this in advance, it will take him a second to decide that Level 2 provides all the information a merchant needs to ship a lamp and to type or say "share 2". The only additional information needed would be the item number. In a work setting, he could transmit "4", perhaps when raising his hand to be considered for an exciting new project. Even while riding his bike, he could submit enough information to be useful.

Matthew's profile is much simpler than the vision aspired to by many companies. It contains no specific product preferences, no history of his online browsing activities, and no preferred companies. But that doesn't mean it couldn't be used to make his life easier. Any merchant located between Matthew's home and office would learn that they could offer him special convenience. Because his information is separated into levels, he can share more information with people and organizations he trusts, but reveal a minimal amount to others. He has 100% control and could decide not to maintain a public profile at all.

If enough people started maintaining this type of profile, it would make widespread personalization much easier while protecting individual privacy. Once companies see that large numbers of individuals are making available their personal preferences in a standardized format, most companies would learn to utilize such information to both their and the individual's benefit.

There's only one challenge with this type of approach: There's no obvious way for a company to profit only by distributing the software that lets Matthew manage his information. All the firms that currently offer similar software make their money not from the individual software – it's typically free – but from helping companies market more effectively to consumers.

YOUpowered is a company trying to balance what it calls permission-based personalization with a service that fosters consumer trust. It exists in the ill-defined territory of privacy intermediaries, where none have yet to turn a profit. Here's the pitch to consumers for YOUpowered's service, called Orby, which helps protect a person's privacy:

- Orby fills in forms with just one click;
- Orby remembers all those passwords;
- Orby makes website-privacy policies easy to understand;
- Orby gives you complete control of your personal information;

The catch is that YOUpowered doesn't make any money from individuals. It generates revenues by selling SmartSense server software to companies, who use the software to track the behaviour of individuals not only at their sites, but at other sites as well. The companies also get access to the profiles that each person creates in Orby. The result is a company that's highly sensitive to privacy concerns, but that considers its most important relationships to be with its corporate clients, not with individuals.

I've yet to see a company that has succeeded by profiting solely from serving individual interests, but I believe this will be the case soon. In traditional businesses, the purest embodiments of this business approach are agents and certain professionals: sports agents, talent agents, and personal injury insurance firms working on contingency. They all profit only if their clients profit. They don't claim to represent an individual, but take revenues from others, as most companies do. David Zimmerman, CTO of YOUpowered, sympathized with the goal but pointed out a potential flaw in the analogy.

"The people who have agents," said Zimmerman, "profit thanks to their extraordinary talents. The agents just help them get value for their talent. What you are talking about is more similar to a union, in which the members are all equal and negotiate based on their collective bargaining power." He conceded, however, that the top 20 or 30% of individuals probably have sufficient buying or negotiating power to enjoy agent-like services, especially as technology makes it increasingly possible to provide such services.

Black Pearl's Hammitt believes it's inevitable that power will shift to the individual level. "People will set up their own agents, who work on their behalf," Hammitt said. "Under this scenario, individuals will be able to create a model of their preferences, controlled by that person, and the agents will search for things of interest to that person." This approach, combined with location services on a mobile phone, enables the service that saved Tony Durkins's Valentine's Day. People will be able to input their brand preferences, sensitivities to price, level of patience, and many other criteria.

There are a number of companies, some in existence, some to be invented, that have the potential to reap immense profits by acting as agents that serve the needs of valuable individuals. I call this approach embracing the individual's agenda. The key is for a company to be able to take a percentage of an individual's financial transactions in return for netting the person more than he or she could have generated on his or her own. Think of an agent who supports a professional's career, feeding him or her customized information and training, alerting the person to job openings that meet a person's highly specific criteria, and managing many of the details of daily work life, all in return for a modest percentage of income. The concept seems absurd today, because no service yet exists that could make a big enough difference to motivate people to sacrifice a portion of their salary. Yet, in some niches, people gladly trade income for services. I routinely give speakers' bureaus a percentage of my income in return for obtaining speaking engagements.

While most businesses can only have contact with people during a limited range of hours, trusted agents will have the right to contact you anytime and anywhere, as long as they are obsessed with your needs. You'd be happy to be awakened by a 4 a.m. phone call from such an agent, if the call informed you that your 6:30 a.m. flight had been cancelled and you had been rebooked on a 9 a.m. flight. But if the agent tried to sell you life insurance during the same call, the relationship would be over.

When thinking about agents, beware of two traps. The first is failure of imagination, and the second, naivety. Software-based agents, churning through vast databases of information, can make individuals literally a million times smarter than they are on their own. Already people can instruct services such as Jamjar to search for cars for sale that meet certain criteria. Expedia offers a service for airline tickets, alerting people when there is a fare sale on their favourite routes. But these services are all centred on certain products, instead of on individuals. The next Richard Branson will build his or her fortune by figuring out how to combine all these services into one. When the same agent can book flights, purchase theatre tickets, buy groceries, do research, order winter boots, and warn a person about traffic on his normal route home, then individuals will have a powerful incentive to share personal information. At the same time, such an agent will be dealing with such a broad range of transactions that the agent will be able to generate substantial profits by taking a relatively small commission on each.

Imagine that a person is unhappy with the seats she holds for a theatre event six weeks hence. She could instruct her agent to search continually for cancellations that would free up two front-row centre seats for any Thursday night performance in the six weeks before the show already reserved. If successful, the agent would automatically switch the seats and inform its client. The more powerful and trustworthy an agent, the more information an individual wants to share.

My wife and I both work, and we have three children, and we are blessed to have a highly capable babysitter who not only cares for the kids during weekdays but also manages many of the details of our household. We are eager to give her as much responsibility as she can handle, and she ends up occasionally dealing with contractors, managing school and community-related issues, and keeping our pets healthy. I feel comfortable she won't sell our personal information to others, so I don't hesitate to give her our credit-card number or the names and addresses of our friends, and even the keys to our cars. But even this remarkable person can't match the potential of agents soon to come. She can't stay up 24 hours a day, 7 days a week, constantly searching hundreds of databases for information, opportunities, or potential problems that might impact our lives.

It's difficult to overstate the profit potential from becoming a trusted agent whose reach extends across most industries. Such an entity could choose to focus on the most affluent fifth of consumers and capture perhaps 70% or more of their total purchases in a year. Since the agent would not take physical possession of goods, it would have no inventory costs and would exist primarily as a data processing and customer-service centre, taking a percentage of each transaction in exchange for tailoring the transactions to meet an individual's needs.

But these people are not just consumers – many are also in the workforce – and thus the agent could manage their career needs as well. The traditional response to this notion is that it's impractical; I've described a scope of activities that span many different industries and types of trans-actions. But in real life, the busiest and most effective people tend to rely on one or two others for such support. A busy entrepreneur often has a capable assistant, and it's not unusual for this person to straddle the border between the entrepreneur's work and personal lives, saving the entrepreneur time and stress whenever possible and practical. Even in our two-career times, some couples decide that one spouse should not work and should instead manage the household. It simply wouldn't be helpful to have eight different personal services that each attempt to manage a different slice of a person's life. In my discussions with others about this concept, most people agree that one to three such relationships would be ideal.

It's all too easy to be naive about the difficulties of building a profitable business capable of surviving strictly as an agent working on behalf of individuals. Nearly every company that attempts to go towards such a path ends up supplementing its income with revenue derived from other corporations, or ends up completely focused on corporations for revenue. Even if the profit challenges can be overcome, significant technical challenges remain. It's hard enough to set up a service that searches airline and hotel databases, as Expedia has. Now imagine doing it with every type of service an individual could want to access. From a practical standpoint, it makes sense for agents to focus on certain industries or services. Unfortunately, from an individual's standpoint, it doesn't. The more organizations a person has to trust with personal data, the greater the odds that one will handle it in an unacceptable or irresponsible manner. Also, the more an agent focuses on certain industries, the more likely it will owe allegiance to the companies that lead that industry, rather than the individuals it serves.

If nothing else, the spectre of trusted agents will motivate companies to balance their interests more evenly with those of individuals, in somewhat the same ways that trade unions did in the earlier stages of our industrial age. To paraphrase what Don Peppers and Martha Rogers have been saying for years, it is far more important for companies to own the relationship with the most valuable stakeholders than to own the assets that produce products.

It's also important to recognize that no computer will ever fully replace the comfort of interacting with people you know and trust. There will always be a need to balance technological solutions with personal knowledge and contacts.

Everyone you know has a public profile, available online. So when you want an opinion, you have numerous choices. You can still go to a public community, such as Epinions, and get recommendations from people you don't know, but who have been rated as trustworthy by others. You can get an opinion from a well-known company, such as Amazon.com, that is motivated primarily by the desire to (a) sell you as many products as possible, and (b) make a profit. Because of (b), such firms might make biased recommendations, because suppliers pay them to promote certain products. Or, you can check out what your friends, neighbours, and peers think, through a system that we'll call Virtual Country Club (VCC).

Through VCCs, you can give other people access to your opinions, ideas, and experiences. You control who gets to see your information. You might set up a rule that says "turn down all requests from companies," so that you wouldn't have to be bothered by marketers who want access to your profile so they can sell you stuff. In the early days of using VCC, you'd send messages to friends and peers asking for access to their profiles. Once approved, you can search and see whatever a person chooses to include in his or her profile. Some people keep their profile on all the time, which means that every product they buy and website they visit gets recorded in the profile. VCC adds ratings based on a person's usage; if you use a website every day, the service ranks it as a favourite.

With this sort of individual control, it becomes easier to know whose opinion you should trust. Suddenly, it's practical to survey friends, peers, and family members on even minor topics, because doing so requires no additional work of your friends. You simply search the information they already have online, and perhaps compare it to results from more corporate services. To broaden your network, you simply ask friends to recommend that their friends allow you into their circle of approved members, in return for the same privilege. To manage your network, you have the capability to sort friends into categories that you define – such as local friends, inner circle, or church members – so you can set different levels of sharing for different groups. At any point, you can change the people or information in your network.

One of the fundamental changes in the way we interact with each other is that we will be comparing with growing frequency opinions, experiences, prices, purchases, performance levels, movements, and ideas. As the obstacles to comparing opinions and analysing data shrink, we will find ourselves asking questions that never before seemed worth the effort. This will change the way we interact with each other and the way we make decisions.

At most companies, employees have little idea how their performance compares to others. Even at firms that label employee performance as "above expectations", people lack specific comparisons. But as data trails multiply and data analysis improves, firms will be able to report a variety of statistics, from how much money an employee's contributions generated for the firm, to how many peers or customers ranked the employee as invaluable. Much of this information will come in the form of feedback from employees, customers, or partners. The more we interact through computers, the easier and less expensive it becomes to collect such feedback. At the same time, corporations will be under growing pressure to do something about the feedback, because people have no motivation to provide feedback unless doing so changes future events.

The challenge is keeping this information accurate. Otherwise, it is worthless, or worse. That's why current online opinion services allow users to rate each others' opinions, which makes it easier for subsequent users to spot unreliable contributions.

A Cautionary Note

Not all personalization is the result of a massive computer sifting through dozens of databases to automatically deliver a certain type of treatment to a person. Most initiatives start with employees who have a good idea. A marketing manager, for example, might have a hunch that people who order diet products from her company would actually be more likely than the average population to respond to a promotion for vitamin pills. Based on this hunch, using new data-analysis software applications that used to be the province of highly trained quantitative analysts but increasingly are accessible to normal business managers, she could sit at her desk and compile a list of 5,000 customers. Within minutes, she could compose an email, and send the offer.

It's both liberating and risky to provide business managers with access to data analysis and mining tools that used to be the exclusive domain of true experts. Web-based access to such software makes it possible for large numbers of managers across a company to get easy access directly into the company's databases. Previously, most managers had to rely on reports created for them by others. These reports were often sanitized to make the data easier to understand.

In 1999, I interviewed Mark Estberg, then senior manager of business applications for Visio, which was installing software from one such firm, E.piphany. Every time a new business unit installed E.piphany, he said, the initial response from business managers and executives was the same: angry emails and phone calls to IT. "They can't believe that the data in our system is so bad, and assume there's a problem with the new system," Estberg said. "But very quickly they realize that E.piphany is showing us for the first time the actual state of our data. E.piphany is working perfectly; it's the underlying data that's flawed or incomplete."

Data analysis and mining tools make it much easier for managers to assess the quality and reliability of information in their databases, which is a good thing. By doing this, they can take corrective actions and better understand when to be sceptical about results. But the tools also make it possible for untrained employees to make flawed judgements, simply because they don't understand statistics or information management nearly as well as real experts do. E.piphany, for example, under the mantra of unlocking the data present in your company, has made it possible for managers in companies such as AT&T, Bank of America, KPMG, Staples and Hewlett-Packard[3] to search multiple databases with relative ease.

On the whole, the opportunities outweigh the dangers, and most companies are cautious about which employees get access to what level of data. My point is simply that we need to spend more time thinking about training people to use technologies that get more powerful every day and to take more care to set up processes that protect both the public good and the corporations that acquire the tools.

Brian Walker had a hunch. There were over 1,500 employees in his division, making an average of £34,000 a year. He knew at least 10%, perhaps more, didn't carry their weight. It was easy to spot half of the underachievers, but difficult to identify the rest. In the past, he would have gone to his team of business analysts and asked them to conduct a study that might produce a means to identify the laggards, but that took weeks, if not months, and he now had direct access to all the firm's databases.

Last weekend, he sketched out a number of assumptions and created bullet points that put his gut instincts into words. Using these notes, he devoted about three hours late on Thursday night to searching multiple databases. He wasted the first 45 minutes trying to figure out how to get the new system to work, and he really didn't understand the statistical processes that produced most of the reports, but eventually he was able to get reasonable answers to his inquiries. Brian was able to generate lists of employees by salaries earned and by bonuses paid, and then compare their compensation to the performance of their business units. He compared people with similar jobs to colleagues in other divisions.

By the time the clock struck 11 p.m., he had confirmed three of his seven assumptions. The only problem was, he didn't have the slightest idea whether his results were right.

Employees may be using tools they don't understand, to produce results that will be formatted in a polished and seemingly credible manner, then presented as the results of the company's new data-analysis system ... but the results may be wrong. Many savvy managers will catch these errors, or at least be prudent enough to check questionable results with trained statisticians. But given the democratization of powerful statistical tools, it's likely that a higher proportion of analysis will be flawed. Black Pearl's Hammitt said, "I hope that managers will still test for integrity in each process and conclusion. Any model could be flawed. I'd like to think most will test where the model is weighting, and where data is correlated." While I agree with her hope, I'm not sure how many managers understand how to interpret her comment, never mind do it.

Hammitt acknowledged that people have an inherent trust of

credible-looking reports. "People have a tendency to stand at attention if they get a report, she said.

This is part of the larger problem of powerful technologies that are now widely available and thus getting into the hands of people who are not ready or responsible enough to use them. In February 2001, the Anna Kournikova virus spread across the world in a matter of hours, infecting over an estimated one million computers. Masquerading as an image of the popular Russian tennis star, it arrived as an email with the subject "Here you have, :o)". Inside was a message that said, "Hi: Check This!" and the attachment, "AnnaKournikova.jpg.vbs". Opening the attachment released the virus, which allegedly was created by a 20-year-old Dutchman, who surrendered to police on Valentine's Day. CNN reported that the man claimed he did not even know how to program computers. Instead, he used a virus tool kit known as a Visual Basic Worm Generator to create the virus.

"It's horrifying," said Mikko Hyponnen, virus research manager at F-Secure Inc.'s European's headquarters, the firm that tracked down the alleged author. "Someone who doesn't know how to program computers can produce a virus that infects hundreds of thousands of computers."[4] Whether or not this suspect truly used a virus generator, the example shows the potential dangers of unleashing advancing technologies.

What's the right way to leverage personalization technologies? We're still years, perhaps decades, away from being able to answer that question with confidence. But it is possible to make reasonable assumptions regarding the difference between meaningful versus superficial personalization, and between needlessly complex versus elegantly simple approaches. By doing so, we can chart a safer course that shows what should be personal.

Thought Exercise . . .

Six Billion Geniuses

Here's the recipe for a world full of geniuses: take thousands of databases, add analytical processing power that's a million times more powerful than anything we had, say, in 1998. Toss in a healthy regard for the preferences of each individual, and a willingness to use the databases and processing power to serve their needs. Combine with out-of-the-box thinking from professionals such as yourself, and mix with interactive communications that give everyone on Earth inexpensive access to everyone and everything else. Let it rise for one to five years. Serves 6 billion.

All you need do is add your own out-of-the-box thinking. What wonderful new uses can you envision for this recipe? Maybe you desire a health-alert network that instantly brings news of medical advances to people who suffer from a related disease, or, for something less lofty, a pager that lets you know when any sporting goods store in the country drops the price of your coveted pair of skis 40% under retail. You might want a chat network that verifies that the woman with whom you are speaking is truly what she claims, and she can trust that you are, too. Wouldn't it be nice to find a news article from months ago that contains information of interest to you, and to be able to find all articles published since on the same subject with just a single command? Think about your personal life, your business, or the needs of people you serve. What marvels can you produce with your own imagination? Chances are the results won't be as far-fetched as you might think.

Notes

1   In the spring of 2001, Net Perceptions went through some hard times. It laid off half of its employees and its market capitalization fell through the floor. Its future was unclear. Still, the company was a pioneer in this space, and we can learn a lot from its experiences. In the spirit of full disclosure, I own stock in both Net Perceptions and E.piphany, which is also mentioned in this book.

2   I went out to dinner with a friend after writing this passage, and before we discussed anything having to do with this book, he mentioned that he had looked up the prices that several of his colleagues had paid for their houses. He was shocked, for example, how little one senior executive had paid for his house, and he searched for the prices of other houses on the street, learning that his colleague lived in a marginal area, despite his wealth and success. "Why," I asked, "did you do this?" His response was simply, "Because I could."

3   All are customers of E.piphany, a leading software company, as listed on 14 February 2001, at
http://epiphany.com/customers/index.html

4   "Kournikova Virus Suspect Arrested," CNN.com, 14 February 2001
http://cnn.com/2001/TECH/internet/02/14/kournikova.virus/index.html

5

Personalize What?

Woman tied to railroad tracks (WOMAN): Help! Help!

Train racing towards woman: Skreeeeech!!!

Police constable (PC): Can I offer you some tea?

WOMAN: No! Untie me! Help me!

PC: Perhaps a biscuit?

WOMAN: It's going to kill me! Hurry! (Minutes later, after the policeman has saved the woman)

WOMAN: Do you really have tea and biscuits?

To paraphrase noted psychologist Abraham Maslow, when we finally get what we want, we want something else. His "Hierarchy of Needs" is one of the most often cited theories of human motivation, and we can use it to think in an organized fashion about the needs of both individuals and corporations. To me, the most important insight in Maslow's theory is that there is a predictable pattern to the order in which people will pursue certain needs. Since personalization revolves around satisfying individual needs, it would be extremely convenient to be able to make predictions with some reasonable accuracy. By doing so, we could figure out what should be personalized, and when.

You're probably already familiar with Maslow's basic ideas. To simplify a bit, he said that starving people don't have the luxury of worrying about whether they prefer French or Japanese food; they just want food. But once people have no problem obtaining food, they start to think about such refinements. Then, they start to think about higher level needs, such as the safety of their children walking to school, or their desire to be part of a community. Maslow initially proposed a "ladder" with five rungs and suggested that people start out on the bottom rung, work to satisfy their basic needs, and, as they do, then turn their attention increasingly to the next rung. He later refined his model to include eight rungs, which I've taken the liberty of relabelling here to better suit our own needs.

While there are obvious differences between people in their needs, Maslow's "ladder" outlines common behavioural tendencies. It works best when you recognize the common human tendency to shift our focus to higher order needs as we become comfortable that we can survive. But be careful about interpreting the ladders too literally. Spiritual individuals, for example, may sacrifice material comforts to help others. The late Mother Teresa skipped the bottom seven levels to focus her life on the eighth one. But this ladder of needs still provides us with a useful way to talk about individual needs and to look at how companies can address them.

Since people and companies are both incredibly complex creatures, we need a quick way to compare the needs of an individual to the capabilities of a company to satisfy that individual. By capabilities, I mean not only the products and services offered, but also the culture, sensitivity, processes, and trustworthiness of a corporation. The Personalization Ladder provides us with a simple mechanism to do this.

This framework is useful for our purposes because it's not organized around specific products or services, as companies are. Too often, company executives make bad decisions about personalization because they are obsessed with selling certain items instead of with accommodating individual needs. It's a good thing to be obsessed in this manner if your goal is to sell more products, but that attitude won't take a company far if its goals are to increase loyalty and profits by locking in the best employees/ customers/suppliers.

As you'll see, we can take a business unit and generalize about the highest level of needs that it consistently satisfies. This status is not a fleeting thing; it is the by-product of many different factors, from capabilities to branding, which is the image of the company in the eyes of its stakeholders. Kelloggs, for example, which makes cereal and other grocery items, is adept at satisfying basic needs. People wouldn't think of approaching the firm for help with self-fulfilment, and, even if they did, the company is utterly unprepared to operate on that level. It lacks the tools and processes necessary to interact with individuals on a personal and highly confidential basis.

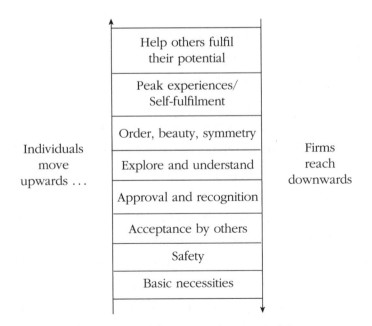

Figure 5.1   The Personalization Ladder

But the inevitable movement towards the personalization of business relationships will impact even firms that meet basic needs, because all firms are racing to lock in stakeholder loyalty, and that race is powered by technology. So all firms will wrestle with issues around what should be personalized and what should remain at arm's length.

In the vast majority of cases, firms should limit their personalization efforts to needs that are at or below the level they already satisfy. It's a wonderful idea to satisfy more of a stakeholder's needs, but it's a lousy – and perhaps dangerous – idea for a company to simultaneously get more personal and to raise the level of needs it attempts to satisfy. This means that, while individuals have a tendency to always move up the ladder, firms must have the discipline to remain focused on the levels they already occupy, and only move downwards, not upwards.

Higher level needs are the most personal. When people seek to satisfy these needs, they are often looking inwards to understand what they were born to do or to examine what the legacy of their life should be. This is the province of religion, dear friends, and human compassion. Many people go through their entire lives without devoting much thought to such needs, mainly because they are struggling to meet their more basic needs. Few large companies are prepared to operate at this level, because doing so requires a bond of confidence and sensitivity. "Spiritual fulfillment, just £500" is not a typical FTSE 100 offering.

As companies start to accommodate personalization, they need to make an accurate assessment regarding the type of needs they are currently able to satisfy. To make an accurate assessment, look at the company from the perspective of whatever stakeholder group the firm is discussing. Especially with large companies, they can be many different things to different people. Customers who purchase nothing more than printer cartridges, for example, may view Hewlett-Packard as satisfying basic needs, but business customers who rely on HP's consulting group may view it as helping support their desire to explore and understand. At the same time, thanks to HP's strong culture, employees may view it as being able to support their highest order needs. I'd plot the three different perspectives in this manner:

HP ... as seen by printer-cartridge buyers	HP ... as seen by consulting clients	HP ... as seen by employees
Help others fulful their potential	Help others fulfil their potential	Help others fulfil their potential
Peak experiences/ Self-fulfilment	Peak experiences/ Self-fulfilment	Peak experiences/ Self-fulfilment
Order, beauty, symmetry	Order, beauty, symmetry	Order, beauty, symmetry
Explore and understand	Explore and understand	Explore and understand
Approval and recognition	Approval and recognition	Approval and recognition
Acceptance by others	Acceptance by others	Acceptance by others
Safety	Safety	Safety
Basic necessities	Basic necessities	Basic necessities

Figure 5.2   Hewlett-Packard

Plotting a company in this manner reveals certain truths about not only its role in a person's eyes – its branding, you might say – but also highlights the probable state of processes and cultural norms within the company. HP has a highly developed culture and well-developed processes that strengthen its relationships with employees. In contrast, until recently HP's printer division didn't even know the identities of customers who purchased printer cartridges. This means that HP is ideally positioned to leverage personalization internally, but that it has a more limited set of opportunities with its printer-cartridge customers. It has far more sophisticated opportunities with its consulting clients, who are accustomed to receiving customized treatment.

To move downwards from any point does not require an immense cultural change at a company. In contrast, moving upwards requires the company to change virtually everything about the way its people, processes, and technologies work. This is a much harder shift, and should only be attempted with eyes wide open and support at every level of the company. Even then, there are great dangers of a misstep, especially where privacy is concerned.

I'm not arguing that companies currently meeting only the lower order needs should ignore opportunities to satisfy higher order needs of their employees or customers. I'm simply saying that they don't need personalization, powered by technology, to accomplish this. A better place to start is with the basic culture and processes of the company. Changes here will make a broader range of personalization possible at a later date.

Peg Neuhauser, a consultant and author of *Culture.com*, argued that "A business strategy based on personalized service requires a personalized culture. If the systems, norms, behaviours, language, and habits inside your organization are not personalized, you will never be effective at providing personalized services to your customers." So if you don't personalize the way you treat employees, you'll never succeed at using the same techniques to lock in the loyalty – or increase the value – of customers.

Neuhauser is cautiously optimistic: "I hope the changes in the way we do business – personalized business strategies – are permanent and will necessitate personalized corporate cultures to deliver on those strategies. We will just have to wait and see what happens in the future. But for now, it is turning to a personalized world of business. There are a lot of people out there who are glad about that, including me."[1]

Here are three quick examples of how different companies could assess their capabilities as perceived by *customers*. To keep this simple, I'm basing these examples on a generalized customer perspective.

Monsoon

This spirited retailer with 135 UK stores and a few dozen more overseas does more than simply provide basic necessities. It helps women and children look and feel great. Here's a tiny bit of copy from its online catalogue, talking about the fact that its current collection "shows a spirit of colourful femininity that has become just a little tougher, prettiness that has grown glamorous."

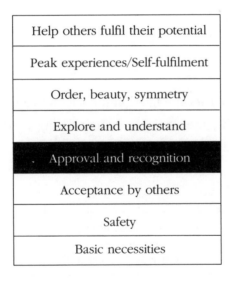

Help others fulfil their potential
Peak experiences/Self-fulfilment
Order, beauty, symmetry
Explore and understand
Approval and recognition
Acceptance by others
Safety
Basic necessities

Figure 5.3   Moonsoon

The implicit message is, the woman who wears Monsoon clothing deserves approval and recognition.

I classify companies by identifying the highest level need that they satisfy as well as all the lower level needs below. In this case, you could argue that Monsoon satisfies a need for order, beauty, and symmetry, but it would be hard to argue that the retailer helps women explore and understand. That's why I put the company at this lower level.

If Monsoon wanted to make its relationship with women more personal, it could credibly migrate to any of the three levels beneath its current position. For example, it could introduce any number of time-saving services designed to help women deal with the basic necessities of life, or it could partner with companies that satisfy such needs. I would not, however, recommend that Monsoon position itself as a provider of customized stock-market research. It has never before collected or managed the type of personal information necessary to support such a service.

Tesco

Tesco is the number 1 food retailer in the UK. While focused on the most basic level of needs, the chain is demonstrating a deft touch at using personal data to the benefit of individuals.

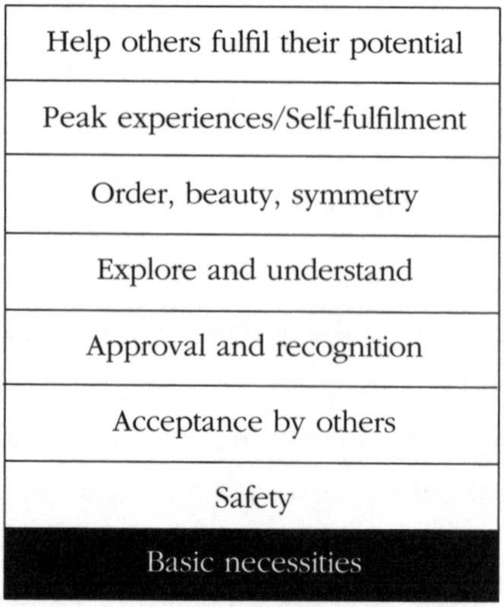

Help others fulfil their potential
Peak experiences/Self-fulfilment
Order, beauty, symmetry
Explore and understand
Approval and recognition
Acceptance by others
Safety
Basic necessities

Figure 5.4   Tesco

Tesco.com describes itself as the largest and most successful Internet-based grocery home-shopping service in the world. In the UK this service is profitable, has almost 1 million registered customers, 70,000 orders per week and annualised sales running at a rate of £300m ($420m).[2]

For years, Tesco has been aggressive in building loyalty among its customers, first through its Club Card programme – which offers customers points in return for each £1 spent – and now through its online shopping services. Today, you can not only buy groceries through Tesco.com, but also DVDs, books, appliances, computers, clothes, flowers, and wine.

Through all these efforts, Tesco is not going upwards, but instead is attempting to extend its franchise horizontally to a greater extent than any of its competitors. This is a strategy designed to lock in customer loyalty, as well as to gain a larger share of each customer's business.

Why, you might ask, isn't Tesco moving upwards? Doesn't its growing assortment of products and services qualify as "explore and understand"? I'd argue not, because the selection still concerns basic human needs. There are some people, such as passionate vegetarians, who consider their diets to be an important component of their quest for self-fulfilment and/or to help others. But these people are in the minority, and Tesco's site does not cater to such needs. Still, you will be able to find times when this approach isn't exact. It's not designed to supply precise answers, but rather to provide a useful framework for thinking about the appropriate level of personalization.

Backroads

Billed as the Number One active travel company, Backroads started as a bicycle-touring company. I first discovered it in the late 1980s, when we spent five days on a Backroads trip through California's wine country. In 1998, my wife and I toured the Canadian Rockies with the company.

Backroads is in the business of providing once-in-a-lifetime adventures, except it makes it easy for people to have such adventures on a regular basis.

Help others fulfil their potential
Peak experiences/Self-fulfilment
Order, beauty, symmetry
Explore and understand
Approval and recognition
Acceptance by others
Safety
Basic necessities

Figure 5.5   Backroads

Recently, we received a postcard from Backroads. The front side was a breathtaking photo taken on our trip, looking across an exquisitely green lake, filled with glacial till, towards the snow-capped peaks. The caption read, "Canadian Rockies: July, 1998." The back referred in detail to our trip and explained that based on our two previous trips, the firm had created a personal Web page just for us, which contained recommendations and details about six other trips we might wish to consider. I was online within minutes, and my wife and I are working out a way to make room for another memorable trip.

For us, despite the infrequency of our trips, Backroads represents one of the best experiences we have ever had with any corporation. The company is talented at managing every detail of extremely complex trips, superb at hiring staff capable of working 16-hour days while being unre-lentingly cheerful, and understands personalization better than almost any firm I can name.

If Backroads chose to stray from its adventure-travel niche, it has plenty of room to migrate downwards. There are lots of reasons why it may not choose to do so – the desire of owner Tom Hale to work outside might be one of the biggest – but the firm already has mastered the skills and sensitivity necessary to make business relationships personal.

Here's one more example, from the perspective of employees.

McDonald's

Entry-level workers at McDonald's are running hard just to get the basic necessities of life, but McDonald's gives them a path to goals at least three levels higher. For employees who embrace the firm's motivational messages, the firm provides a clearly defined path up through management of individual units, into district management, and then on to either ownership or corporate management. Not every employee takes this path, but, without it, the firm would have a much weaker grip on the quality of its operations.

McDonald's could leverage personalization to make the management track more compelling by better charting an employee's progress and better outlining additional steps employees need to take to reach the next level. On the other hand, McDonald's would have a difficult time offering to meet an employee's desire to "explore and understand". Doing so would divert employees from the discipline of focusing on the basic operations of a fast-food restaurant, and it wouldn't sound credible coming from McDonald's, which serves up hamburgers, not knowledge.

To use the Personalization Ladder for guidance in understanding the level of personalization desired by an individual, use it to plot their current state, just as we did with corporations. But this time, we are looking at

Help others fulfil their potential
Peak experiences/Self-fulfilment
Order, beauty, symmetry
Explore and understand
Approval and recognition
Acceptance by others
Safety
Basic necessities

Figure 5.6   McDonald's

individuals. Their position on the ladder indicates the type of personalization they are most likely to value. When a person can't even meet her basic needs, she isn't likely to value a personalized shopping experience. In all honesty, she's probably too worried about food and shelter to even become aware of such loftier services. But later on, when she becomes more comfortable and starts to seek higher order needs such as acceptance by others, she has the breathing room to consider such previous "luxuries". These luxuries are especially attractive when they come at little or no additional cost to the person.

In the hypothetical case of Catherine Leslie, a successful professional, she is goal oriented and strongly focused on reaching the next management level. Not interested in fuzzy management concepts or speculative exploration, she wants to book new business and show a hefty profit at the end of the year. She's likely to be most receptive to personalization that removes distractions – that is, minimizes the time needed to meet her basic needs – and that helps her maintain a positive reputation among her peers, while doing everything possible to get recognition from the senior-level executive who controls her advancement.

Figure 5.7 Catherine Leslie

The hierarchy-of-needs framework is useful, but only indicates the broad range of needs a company will be able to meet or an individual likely to find valuable. It still doesn't tell us what to personalize. Both Tesco and Backroads start by remembering information for a person, and then use this information to deliver unique benefits to that person. The measure of such benefits is whether each company, thanks to its unique knowledge of the person's needs, could be the only firm capable of meeting them. Backroads alone not only knew which trip we took, but also kept a photograph from the specific trip ... as opposed to a generic photo or a random trip.

The more literally a company uses a customer's information to provide unique benefits, the more likely that personalization will influence a person's behaviour. Most companies ignore or minimize this second step,

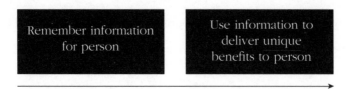

Figure 5.8   Companies and Information

which puts them in dangerous territory with regards to privacy and trust. In fact, the practice of collecting personal information without providing unique and meaningful benefits to that individual is what scares people and motivates legislators to enact stricter privacy laws.

Here are the primary benefits of personalization, with examples of each:

Save time   Eliminate repetitive tasks; remember transactional details; recognize habits and shorten the path to engage in such habits (example: frequently called numbers on a phone should automatically go into the phone's memory).

Save money   Prevent redundant work (example: make it easier for employees and suppliers to know someone else has already solved the problem that they are currently facing); eliminate service components unnecessary to a person; identify lower cost solutions that meet all other specifications.

Better information   Provide training; filter out information not relevant to a person; provide more specific information that is increasingly relevant to a person's interests; increase the reliability of information; replace "average" information with information specific to that person's environment.

Address ongoing needs, challenges, or opportunities   Provide one-stop services; allow flexibility in work hours, job responsibilities, and benefits; accommodate unique personal preferences (example: allow employees to customize their office space, within certain boundaries); recognize and reward achievement with special treatment.

Jeff Bezos, Amazon.com founder, has said, "To be nine times bigger than your nearest competitor, you actually only have to be 10% better." In Amazon's case, better means being the first to foster a Learning Relationship and continuing to be at least 10% ahead of the competition. This is enough to keep a Learning Relationship intact. If you go through the above list, Amazon.com is addressing each point:

- Save time: one-click ordering.
- Save money: up to 40%.
- Better information: BookMatcher service remembers your preferences; it can tell you other books you're likely to enjoy based on purchases and/or preferences of similar customers.
- Address ongoing needs, challenges, or opportunities: they now offer a host of other consumer goods as well as books.

If these are the individual benefits of personalization, how can a firm deliver them? All personalization has its roots in the abilities to gather, filter, and sort information, and there are certain information-driven activities that deliver one or more of these benefits. Here are 11 ways to make it personal:

Combine   Merge information a person already has with that of others, to provide additional insights. For example, one researcher has studied the side effects of a certain drug, while another understands in detail the mechanisms that cause the drug to work. Together, these two bodies of knowledge have much greater value.

Compare   Show how prices, quality, or specifications of one option match up to others. This type of comparison shopping is already in great evidence online.

Connect   In most large companies, data exist in "silos". Information about a customer or employee might exist in eight different databases, in different parts of the firm. Disconnected, this data can't help the company or the stakeholder. Firms can connect this data, providing a more accurate picture of the firm's interactions with that person. The flip side of this activity is that connecting previous disparate data removes a level of privacy and enables the company to learn things about a person that it was never able to deduce before.

Explain   Clarify how, when, or why to use a product or service, or to perform a task, precisely when a person needs such help. People need easier-to-understand-and-access guidance about navigating new technologies, processes, and challenges.

Find   Locate a person, product, or service based on supplied specifications. This answers the question, "Who can give me what I need?"

Monitor   Track the status of events, news, or actions of others. While customized news and information is already popular online, most firms have barely scratched the surface of monitoring possibilities. For example, some day-care centres now provide parents with Web access to video cameras that allow them to see what their children are doing at the centre.

Recommend   Suggest a course of action based on historical data, the current environment, or predictive models. The greater the consequences of a bad decision, the greater the value of trustworthy recommendations.

Remember   Most people are still more frustrated about what companies forget about them than what they remember. Firms can strengthen relationships simply by remembering what happened the last time they interacted with an individual, and not requiring the person to tell them the same thing twice.

Reveal Highlight a pattern or conclusion that was not previously evident. By showing an employee non-obvious differences between the composition of their rejected vs. their accepted proposals, a firm could simultaneously increase efficiency and reduce stress.

Sort Change the order or grouping of information, making it easier for people to see patterns. Stop & Shop does this by re-sorting consumer purchases every time a consumer chooses a different objective (for example, low sodium vs. high fibre).

Trigger Prompt an action when certain criteria are met, such as the purchase of an item when its price falls below $150. This is the essence of the trusted-agent concept, using computer processing power to serve as a constant watchdog for the person's interests.

It is time to tie together everything we've been discussing. To help you imagine the possibilities, or perhaps power a brainstorming session at your firm, let's take the four main benefits of personalization and map them next to the eleven activities we just reviewed. The result is seen in Figure 5.9.

In my work with companies, we'll sometimes choose a particular stakeholder group that we consider to be currently focused on a certain level of needs, and then go through each activity, brainstorming ways that we could develop a service that would deliver one or more of the four benefits to the stakeholders. Once we generate a healthy list, we go back a few times, first to rank the likely value to the individuals we wish to please, and later to assess the costs and difficulty of providing the service. Finally, we consider the likely benefits to the business. The ideas that survive provide win–win benefits for all parties, at a cost and level of effort that is practical, given the business unit's circumstances.

Imagine that I'm working with a successful entrepreneurial company, which has now been public for a few years. Many of the most talented

	Benefits to individuals			
Activity	Save time	Save money	Better information	Address ongoing needs
Combine				
Compare				
Connect				
Explain				
Find				
Monitor				
Recommend				
Remember				
Reveal				
Sort				
Trigger				

Figure 5.9 Benefits to individuals.

employees are worth millions, at least on paper, and are suddenly at a time in their lives where their attention turns to higher order needs, such as peak experience and self-fulfilment. While there is certainly a point at which no company can accommodate all of an employee's higher order needs, it's worth my client's time and money to retain as many of these employees as possible, and more money won't do the trick. Instead, we use this approach to look for ways to leverage personalization to give these employees more freedom (that is, save time and get better information, faster.) We also try to increase the perceived value of being part of the company's community, by connecting them with other employees or partners who share similar, or complementary, talents and goals. The goal is to lock in the loyalty of employees even when they no longer need to work for a living, and that takes a personal approach.

But wait a minute. Halfway through a book on personalization, I just gave you an example that described a group of people with common needs, instead of a single person with unique needs. Doesn't that contradict the whole point of our discussion? Not at all. With the exception of serving extraordinarily valuable stakeholders – an investor who contributes £10 million, or a customer who supplies 30% of your revenues – most personalization starts by identifying a group of individuals with certain needs in common. Collectively, such a group represents two qualities. First, its total value is significant to the firm and thus justifies reasonable investment. Second, its members have needs that are different from other individuals and thus could be better satisfied via a customized approach. Once a firm identifies a group such as this, it can then invest in the capabilities necessary to offer meaningful personalization. Only at this point can a company actually deliver personalization. So first define a segment, then individualize the treatment of each person within the segment.

"No Thanks"

No matter how remarkable or laudable a company's efforts at personalization, there will always be some people who simply are not interested. Whether a firm adopts an opt-in or opt-out approach, it must be prepared to recognize and instantly accommodate any of the motivating factors that would cause a person to decide he or she doesn't want any sort of personalization.

From an individual's perspective, there are many situations or attitudes that make personalization unwelcome ...

Anonymity preferred

There are many reasons why people might not want to be identified, from the innocent – it's a birthday present they don't want their spouse to discover in advance on their credit-card statement – to the unethical or illegal. Some people are simply private, and prefer to mind their own business and let you mind yours. Others recognize the growing infringements on private space and choose to take the cautious route. A. Michael Froomkin, associate professor at the University of Miami School of Law, wrote, "Anonymity may be the primary tool available to citizens to combat the compilation and analysis of personal profile data, although data-protection laws also may have some effect. The existence of profiling databases, whether in corporate or public hands, may severely constrict the economic and possibly even the political freedoms of the persons profiled; although profiling may not necessarily change the amount of actual data in existence about a person, organizing the data into easily searchable form reduces her effective privacy by permitting 'data mining' and correlations that were previously impossible."[3]

Froomkin also pointed out that not every scholar of law believes in the right to anonymity. US Supreme Court Justice Antonin Scalia wrote a dissenting opinion[4] in which he said anonymity is generally dishonourable. "It facilitates wrong by eliminating accountability, which is ordinarily the very purpose of the anonymity," Scalia wrote.

Lack of relevance

People do not want a relationship with companies that have no relevance to them. Computer programmers have no interest in getting to know an executive recruiter who only places sales executives. Homeowners who only buy the finest products for their home will not be interested in a cut-rate furniture store. If you've never been to Scotland, never plan to go there, and don't know anyone there, you don't want to be on the mailing list of the Scottish Tourism Board. On the Web, companies constantly ignore this factor and ask individuals for information before demonstrating to the person's satisfaction that their services are relevant. The prime example is companies that insist people fill out a lengthy form before they can gain access to a demo or to additional information. If a company asks people for information before it has demonstrated relevance, between 30 and 50% – depending on which statistics you believe – will lie to prevent revealing personal information. Catherine Legge, writing in her Net Effects column, nicely summed up a common attitude:

I lie. And, I don't feel guilt or remorse. When it comes to giving out personal information online, I have the morality of Satan's spawn.

Sometimes I'm Candice and sometimes I go by my soap opera diva name, Ms Styles.

I usually live in Beverly Hills because I know the zip code is 90210. When asked about income, I am a student who makes $0 to $12,000 a year.

Lying online is not wrong. It's survival.

The whole utopia of this free exchange of ideas has been choked out by opportunistic e-marketing types. I'm getting groped for information, flashed by profiled banner ads and then some page full of blanks wants to know my favourite colour?

On the other hand, people won't lie when they see relevance. If a firm wants my name and address so it can ship the windsurfer I'm eager to sail next weekend, I'll gladly give it and volunteer that they better jot down my phone number, just in case.

Untrustworthy

If you don't trust a company, it becomes a relationship of last resort. Unless you have no choice, you don't want to deal with it. People don't need proof that a company deserves to be in this category. Often, a small suggestion that this might be the case is enough to justify caution.

Hendrick A. Verfaillie, CEO of Monsanto, believes this is a time when a "shift in society – a shift that started perhaps 40 years ago – is approaching full maturity. That shift has been a movement from a 'trust me' society to a 'show me' society. We don't trust government – and thus government rulemaking and regulation is suspect. We don't trust companies – or the new technologies they introduce into the marketplace. We don't trust the media – or the news they bring us each day. And so it goes with all institutions."[5]

A few years ago, before the Web armed prospective buyers with armfuls of pricing and quality information, anyone walking into a dealership to negotiate for a new car was taught why she should not trust the dealer. The salesperson talks about how high the dealer's costs are and starts to negotiate with the buyer. The salesperson then inevitably has to go and present an offer to his boss, who never actually enters the same room with the buyer. Three minutes pass. The salesperson returns, shaking his head. "He says we need to get at least £15,000, or we'll lose our shirts." The whole process is designed to put the buyer in a weak position and to get the dealer the best margin possible.

A few weeks ago, I bought a car after extensive research and long-distance discussions with a range of dealers. When I picked up the car, the salesperson said to me, "You got a really good deal. I just sold the same car to someone last week for £1,200 more than you paid." This made me feel good but taught me never to trust that dealer. Either he charged the other guy more simply because he could get away with it, or my deal wasn't actually so good, and he was deceiving me.

This lack of trust can't help but encourage the growing tendency of British consumers to buy cars through alternative channels such as telephone and online services. They have no reason to be loyal to dealers more interested in their own good than the buyer's.

Lack of security

Good intentions aren't enough. If a company fails to protect its assets, and those of its stakeholders, then people will not be willing to share anything of value with the firm. Security is like sausage making ... the more you know about it, the less likely you are to be comfortable. People have real reasons to fear that today's centralized networks aren't secure, because they aren't.

In January 2001, the names, email addresses, phone numbers, and addresses of up to 51,000 customers were exposed on the Travelocity website. The customers had participated in two recent promotions, and the lists were accidentally left on a server after the company transferred the computer from San Francisco to New Orleans.[6] The names were exposed for more than a month, said Jim Marsicano, executive vice-president of sales and service for Travelocity. Blaming the problem on human error, Marsicano stressed that no customer order information was compromised by the security hole.

Such events led Scott Culp, of Microsoft's Security Response Center, to develop the Ten Immutable Laws of Security[7] (see Figure 5.10). Note how many of them have to do with people, instead of just technology.

Technology firms are working to solve security problems, although most admit that security is a process, not a single technological solution. Protegrity, for example, takes a granular approach to information security. Instead of simply building "walls" around servers or hard drives, its Secure.Data solution builds a protective layer of encryption around individual data items or objects. Even if someone breaks into a network and steals your personal data, theoretically he can't read it, because each piece of data is encrypted.

Of course, there's a difference between actual security and perceived security. Companies not only have to maintain high levels of security, but they also must convince their stakeholders they are behaving responsibly.

Law #1:	If a bad guy can persuade you to run his program on your computer, it's not your computer anymore.
Law #2:	If a bad guy can alter the operating system on your computer, it's not your computer anymore.
Law #3:	If a bad guy has unrestricted physical access to your computer, it's not your computer anymore.
Law #4:	If you allow a bad guy to upload programs to your website, it's not your website anymore.
Law #5:	Weak passwords trump strong security.
Law #6:	A machine is only as secure as the administrator is trustworthy.
Law #7:	Encrypted data is only as secure as the decryption key.
Law #8:	An out-of-date virus scanner is only marginally better than no virus scanner at all.
Law #9:	Absolute anonymity isn't practical, in real life or on the web.

Figure 5.10   The Ten Immutable Laws of Security

Impossible

Sometimes, people just aren't able to take advantage of attractive offers. If a company, local government, spouse, or neighbourhood forbids a person from moving forward, that's life. Likewise, if people lack the ability to accept personalization – perhaps they lack a sophisticated enough mobile phone, or a fast enough Web connection – it won't happen.

Charles Dwyer, director for the Aresty Institute's Managing People Program at the Wharton School, spends much of his time helping executives learn how to manage without authority. He argues that in today's business environment, reliance on authority is becoming less available and less effective. Instead, managers must learn to influence the behaviour of others, rather than dictate it. This is a good way to think about motivating individuals to appreciate personalization, which is something firms can't dictate without crossing a line into unethical behaviour. But

Dwyer pointed out that there is one key hurdle to leap before anyone can successfully influence the behaviour of another: Can the person do what you want? Dwyer said, "There's really only one constraint on human influence, and that's if the person you want to influence doesn't have the potential to do it. Actually, that's an extraordinarily mild constraint. Most of the time, people can already do – or can be brought to the point of wanting to do – what you want them to do."

Tom Kelly is the general manager of IDEO, a design firm that helped create the first Apple mouse, Polaroid's I-Zone instant camera, and the Palm V Pilot. In a radio interview, Kelly talked about designing the Crest Neat Squeeze standup toothpaste tube. IDEO's goal was to get rid of the messy residue that fills the threads of screw-on toothpaste caps. The design firm came up with a solution that eliminated the threads altogether, seemingly a much better solution. The designers then watched consumers use the new cap and realized there was a big problem. After decades of opening and closing toothpaste containers, consumers had learned that toothpaste caps screw off. When confronted with one that didn't, they didn't know how to open it. Without observing people, it's easy to miss that they are simply not capable of doing all the tasks a company expects them to do.

Infrequent contact

People will have little interest in establishing a relationship with a cab driver in a city they rarely visit, or with the company that installs their new septic tank (a once-in-25-years event). Companies get around this limitation by broadening their services to increase the frequency of contacts. Hewlett-Packard's printer division used to focus on selling printers; now the firm realizes it can make more money selling printer cartridges, as well as paper, and in the process increase the frequency of its interactions with customers.

OnStar addresses this problem for General Motors, which *never* interacted with the consumers that bought its cars. The system, built into many GM cars, combines global positioning with wireless technologies to deliver personalized services. Real people back up the technology, offering 24/7/365 service, accessible at the touch of one button.

The company offers an increasingly broadening range of OnStar services that drivers can access from their cars. OnStar can direct you to over 90,000 hotels and resorts, 500,000 restaurants, 90,000 cash machines, and 70,000 service stations. Personal Calling provides drivers with built-in voice recognition systems that allow them to call a number by saying the numbers. Virtual Advisor provides customers with up-to-date personalized information such as news, weather, stocks, sports, and email, accessible from your vehicle ... all by speaking simple voice commands. The system even provides concierge services.

Little value placed on potential benefits

People may not recognize the value in offered personalization, such as when firms offer to customize product offers. Many people don't want to receive any such offers, period. Employees who are offered personalized training may not value it if they were unimpressed with their previous experiences with the training unit, and thus believe that even personalization won't make the time invested worthwhile.

Anyone who works in a company or department for long enough tends to lose a bit of perspective around this issue. Early in my career, I raised funding for PBS radio and television programs, working at WGBH/Boston. After a while in the job, doing on-air and in-person fund-raising pleas, I couldn't conceive why anyone wouldn't agree that WGBH was a national treasure worthy of a substantial contribution. This type of semi-delusional attitude is what gets people past repeated objections, and it certainly worked for us. It was only years later, after leaving the station, that it dawned on me how low public broadcasting is on most people's list of causes to support, well behind religious institutions, alma maters, and health-related causes.

As companies move towards the personal – and the number of inter-actions increases – it's important to gain greater objectivity about the attrac-tiveness of a firm's offers. Today, in the early stages of our shift towards increasingly personal business relationships, most personalization is still superficial, and way too much of it is mainly personalized marketing. No matter how targeted advertising becomes, it still won't be anything more than a means to an end, and too much of it is flat-out annoying. People tolerate occasional annoyances, but when annoyances multiply, they begin to reek of harassment.

Even highly attractive offers won't make a difference to a person who doesn't value the potential benefits. Think about a new knowledge-management system that theoretically delivers "better" information by filtering out "less relevant" citations. Many researchers may cringe at the thought, because they succeed by looking at raw data and thus understand-ing at a deeper level the background and related elements of a given situation.

Is the Company Ready?

Contrary to widespread perceptions, personalization isn't simply a strategy that a firm decides to embrace. It is a tendency that seeps into numerous systems and processes as an inevitable result of doing business in a wired world. Just as a boat drifts with the current, so, too, are firms moving towards increasingly personal relationships, even before they consciously decide to do so. Any firm that collects information about individuals is in the early stages of personalization. It used to be that firms needed to acquire expensive software applications specifically designed to enable personalization, such as many of the systems we discussed in the previous chapter.

But elements of personalization are creeping into most software programs, including mainstream offerings from Microsoft, Oracle, and Sun. But while personalization is increasing everywhere, there comes a point at which companies need to make personalization a central part of their corporate strategy, as opposed to something that simply requires awareness and sensitivity. Just as individuals change their focus as they satisfy increasingly higher orders of needs, so do companies. Firms can't make business personal until they have established stable customer and supplier relationships and can execute the basics well. Using the same format as the Personalization Ladder, let's take a quick look at how we can visualize the changing focus of a corporation. To do that, I've replaced Maslow's rungs with needs that are specific to the evolution of a company.

At the bottom of the ladder, companies are focusing on building their business. Think of a new company that has a good idea, perhaps has a new product, but hasn't yet expanded its services. Such a company is focused mainly on developing capabilities that others value, whether they are products or services. Once the company does this, the challenge is to build a customer base by taking the capabilities it has and expanding them to the point that the company has a viable business.

In the early stages of a company's life, it's challenging enough simply to be able to meet the quality standards that clients have, especially as the firm grows. Many new firms can produce products in small batches, but have difficulty when demand increases dramatically. But there reaches a point, about halfway up the ladder, when a company has plenty of satisfied customers who are comfortable both with the firm's quality and prices, and the final step to profitability is for the firm to reduce its costs.

It is at this point that it is practical for companies to start to think about personalization, because they have reached a stage where they can compete on quality and on price, but realize what all established firms know today, that neither quality or price gives a sustainable competitive advantage any more. This is one of the main reasons why companies need to make business personal, because by doing so firms can create a Learning Relationship that locks in both customer loyalty and the loyalty of other key stakeholders.

Through personalization and other tactics, companies now start to focus on building share of customer, which means satisfying more needs of the customers they already have. By doing more for the existing customers, companies increase both profits and revenues. No matter what department you work in, every company needs to be driven by customers, and there comes a time when building share of customer has to be the defining goal of the company ... but it is only after the firm has mastered these other challenges. Personalization helps tremendously at this stage because the whole process of personalization says let us learn better what a customer needs and then do something about it.

Beyond this point, firms can think about using personalization to improve their internal operations, to reduce the redundancies that happen in any growing or large company, and also to encourage reuse of innovations and capabilities in a manner that generates incremental revenue without proportional increasing expenses. Firms can expand services and start new businesses by taking work that has already been done and finding new ways to leverage it.

As companies near the top of the ladder, the ultimate goal of personalization is to be able to establish effective feedback loops, which means they are able to listen to key stakeholders and customers and do something different as a result. Changes in feedback should cause changes in behaviour, not just with regards to the company but also with regards to its key stakeholders. If the company does something different or says something different to a stakeholder, then the stakeholder's behaviour should change, too.

In the future, the highest order goal of a company will be to become a trusted agent, to be perceived by key stakeholders as the one entity that is focused 100% on the stakeholder's needs. The stakeholder trusts the company to provide what the stakeholder needs and is willing to pay a fair value for this attention. This is the essence of a non-zero relationship, and the ideal result of personalization.

Thought Exercise ...

What Rung?

Try ranking your business unit on the Corporate Needs Ladder shown Figure 5.11. How far up the ladder has your unit climbed? If you work in a medium or large company, how does your unit compare to the company as a whole? Are you ahead or behind? Are both in a position to focus on personalization? Once you've collected your thoughts, try passing a copy of this page around to a few colleagues – or better yet, buy them a copy of this book – and ask them to take on the same question. You may be surprised to find you have different answers. The interesting learning will come from understanding why this is so.

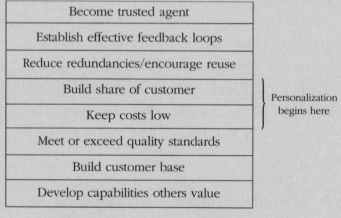

Become trusted agent
Establish effective feedback loops
Reduce redundancies/encourage reuse
Build share of customer
Keep costs low
Meet or exceed quality standards
Build customer base
Develop capabilities others value

Personalization begins here

Figure 5.11   The Corporate Needs Ladder

Notes

1   "If You Want Personalization in the Marketplace, You'd Better Personalize
    Your Company's Culture," by Peg C. Neuhauser, author of *Culture.com*.
    http://personalization.com/soapbox/contributions/neuhauser.asp

2   Tesco press release dated 25 June 2001, "Grocery Works to Implement
    Tesco.com's Store-based Grocery Home-shopping Service in the US
    through Safeway Inc's Stores.

3   "Flood Control on the Information Ocean: Living With Anonymity, Digital
    Cash, and Distributed Databases," by A. Michael Froomkin, *University of
    Pittsburgh Journal of Law and Commerce* **395** (1996).

4   Dissenting in *McIntyre vs. Ohio Elections Commission*.

5   Remarks at the *Farm Journal Conference, Washington, DC, 27 November
    2000*.
    http://www.biotech-info.net/new_Monsanto.html

6   "Travelocity Exposes Customer Information," by Troy Wolverton, CNET
    News.com, 22 January 2001.

7   October 2000, Copyright 2001 Microsoft.
    http://www.microsoft.com/technet/security/10imlaws.asp

6

Unintended Consequences

T wo lessons emerge loud and clear from corporate efforts so far to leverage personalization. First, personalization makes business dramatically more complex. It may be possible – and ultimately more profitable – to do something different for each person, but it's certainly not simpler.

Second, while new technologies are being introduced at a rapid pace, lagging far behind are much-needed changes in the processes, compensation systems, cultures, and expertise levels of most companies. Complex systems already have a tendency to produce unexpected results. Factor in the lag in necessary changes, and you get a situation guaranteed to produce unintended consequences.

For every company that thinks of itself as trying to implement personalization initiatives, there are hundreds more who never use that word but still wrestle with the same forces. Under the guise of digitization, integration, e-commerce, and countless other *termes du jour*, firms are struggling to tear down the walls within corporations, so that Web interfaces can communicate directly not only with legacy systems, but also with systems running at partner and supplier firms. Whether an individual buys a Palm

Pilot or a corporation spends £1 million on a new software application, technologies that people don't fully understand are everywhere. Almost all of these technologies make increased personalization inevitable, and few people have the time to consider the negative repercussions that could result.

Environmentalists use this story to illustrate the dangers of unintended consequences:

In the early 1950s, the Dayak people of Borneo suffered from malaria. The World Health Organization had a solution: it sprayed large amounts of DDT to kill the mosquitoes that carried the malaria. The mosquitoes died; the malaria declined; so far, so good. But there were side effects. Among the first was that the roofs of people's houses began to fall down on their heads. It seemed that the DDT was also killing a parasitic wasp that had previously controlled thatch-eating caterpillars. Worse, the DDT-poisoned insects were eaten by geckos, which were eaten by cats. The cats started to die, the rats flourished, and the people were threatened by potential outbreaks of typhus and plague. To cope with these problems, which it had itself created, the World Health Organization was obliged to *parachute 14,000 live cats into Borneo.*[1]

In an increasingly interconnected world, even small changes have unexpected repercussions. We'll focus in this chapter on the implications of this growing complexity, detail the types of changes that are most urgently needed within most companies, and review the most potentially troubling consequences. We'll be focused on elements at work within a company, but it's important to note that pending changes in privacy laws could also produce such repercussions. In protecting people from unwarranted intrusions, legislators need to be careful not to inadvertently prohibit services that the majority values. Corporate lobbying efforts have a tendency to highlight such potential problems, often citing unintended consequences such as reduced levels of competition, higher prices, and less selection.

Complexity

Complexity is the enemy of profits. Our world is one in which company market values can double or be cut in half in a few days, simply because a firm's performance varies from analyst expectations. Our free-market system puts a premium on predictability, and complexity makes predictability nearly impossible.

What's wrong with complexity? Nearly everything. Complex systems are:

- more likely to fail;
- prone to exhibiting unpredictable and/or unintended behaviour;
- less secure;
- difficult for users to understand and trust;
- difficult to maintain;
- too expensive;
- unlikely to produce the promised return on investment.

We'll examine each of these liabilities in turn, but first you should understand why personalization and complexity go hand in hand. Modularity, which means dividing a system into separate but interconnected pieces, makes personalization possible. The more modules you create, the better able you are to provide unique solutions and services. Companies can divide almost anything into modules: benefit plans, communications programmes, price lists, training programmes, transportation systems, software code, office parties, job descriptions. They can create modules to accommodate the needs of people in different functional areas, geographies, or business units.

Here's the paradox: the more modules a business creates, the more flexibility it enjoys . . . but the more modules, the more complexity.

To illustrate this, let's examine an imaginary benefits programme divided into the following modules. The numbers in parentheses indicate how many choices are available for each option:

- facility location (choose one of five);
- dental option (choose one of three, including "none" as an option);
- add family and household members to plan (four options);
- cash-back option (five choices designed to accommodate employees with no children or families);
- education reimbursement (two options);
- out-of-plan treatment options (three choices).

To determine the number of possible unique combinations, you multiply the number of options together. In this case, that means $5 \times 3 \times 4 \times 5 \times 2 \times 3 = 1,800$ combinations. This is a relatively small number. If you look at the choices on a computer manufacturer's online configurator, you'll find that the combinations number in the millions.

How many situations require over a million options? Are people really that different? In most – not all – cases, the answer is no, and the challenge is to determine the right balance between choice and complexity.

It's not just the number of modules available, it's also how they are presented and used. Whether there are 1,800 or one million possible combinations, suddenly there are a lot more choices than ever before, and a multitude of ways that these choices can be expressed, perceived, and put into practice. That creates numerous changes in the processes that exist within and between firms.

Even a small change in a single process can make a big difference. From watching television, we all know how police line-ups work, right? Six suspects walk into a room simultaneously and the victim tries to pick out the culprit. The problem with this approach is that it encourages the victim to pick someone – anyone – and that this subtle pressure sometimes results in a false positive; the victim picks someone other than the culprit.

A far better result comes from a subtle change in this approach. Instead of having the suspects come in simultaneously, it turns out that suspects should be shown to the victim in sequential order. One at a time, the victim can consider each suspect. This approach does not reduce the number of instances that the victim identifies the actual culprit, but it significantly reduces the number of mistaken IDs. In one study,[2] the simultaneous approach produced nearly three times the false identifications as the sequential approach. So the question isn't as simple as, "Do we have the criminal?" The hard question is, "Do we know how to determine if we have the criminal?"

Firms need to think carefully about ways that they can simplify the process of personalization, and thus increase both its safety and effectiveness. To help make that case, let's examine each of the main problems with complexity.

Tendency to fail

The more complex a system, the more likely it is to fail. I've had the same hammer for 10 years, and it has never failed to work. I've owned computers for over 15 years, and not one has failed to crash, die suddenly, or generally annoy me.

There's a simple reason for this: the more complex a system, the more things that can go wrong with it. Some software manufacturers say their software has become so sophisticated that it can fix itself, but they really mean in certain limited, previously anticipated situations.

When complex systems develop problems, they often require complex solutions. Cars are a good example of this. When your car didn't start, some of us used to be able to lift the hood and find the problem. Today, many cars have computerized diagnostic systems. This makes it faster for authorized dealers to diagnose and fix your car's problem, but also makes it nearly impossible to service the car without investing in an expensive diagnostic computer.

When you consider that nearly every industry is cyclical, and that budgets are tighter some years than others, do you want to burden yourself with a system that requires an expensive and complex operation just to keep running?

Unexpected or unpredictable behaviour

Stock markets are complex systems. Vast sums of money have been invested in predicting the behaviour of such markets, in most cases with negligible results at best. Investment banks have invested hundreds of millions of dollars in an effort to simply be right 51% of the time, and yet no matter what approach they take, their crystal balls tend to function only occasionally.

Much of the unpredictability stems not from technology, but from people. I once observed an order-management system in an IBM call centre that was designed to record the source of every order. To do so, the call operative had to ask the caller to look for a code on the back of the IBM catalogue. In theory, the system should have worked perfectly. In reality, only 10% of the orders, or less, was properly attributed. The first problem was that callers often didn't have the catalogue when they called; they had left it at home or in someone else's office. The second problem was that operatives, being under time pressure to handle as many calls as possible, found it faster to enter a default code (that is, source code = "unknown") than to waste 20 seconds asking the caller to find the number.

In this situation, is it the fault of the system, the caller, the operative, or the management team that failed to properly motivate the operative to record accurately the source code? While what was unexpected was the behaviour of the operative, I'd argue the overall complexity of the system is at fault. If it was designed properly, the operative wouldn't have to enter the number, because both the customer names and the right source code would already be in the database. After all, someone mailed the catalogue to the caller in the first place. Yes, there will be times when catalogues are passed to friends and peers – so the name won't be in the system – but it's easier to deal with a few exceptions than to require an extra step for every single order.

Less secure

Complex systems are far less secure than simple systems. In his excellent book, *Secrets & Lies: Digital Security in a Networked World*,[3] Bruce Schneier talked about the increasing complexity of technological systems: "As a consumer, I think this complexity is great. There are more choices, more options, more things I can do. As a security professional, I think it's terrifying. Complexity is the worst enemy of security. This has been true since the beginning of computers, and is likely to be true for the foreseeable future. And as cyberspace continues to get more complex, it will continue to get less secure."

Schneier cited four reasons why systems will be less secure as they become more complex:

- the number of security flaws goes up as software complexity increases;
- modularity itself is an enemy of security, because security often fails where modular systems meet;
- complexity leads to the interconnecting of systems, which makes it easier for simple problems to get out of hand;
- the more complex a system, the more likely the people running it don't understand it. This makes it easier for them to be taken advantage of by someone who does.

Table 6.1. Increasing complexity of operating systems.

Operating system	Year	Lines of code
Windows 3.1	1992	3 million
Windows NT	1992	4 million
Windows 95	1995	15 million
Windows NT 4.0	1996	16.5 million
Windows 98	1998	18 million
Windows 2000	2000	35–60 million (est.)

I'd highly recommend Schneier's book – it's easy to read yet extremely detailed – because a detailed discussion of security goes way beyond the scope of my book. To reinforce the trend towards complexity, Table 6.1 is a chart Schneier created that shows how Microsoft operating systems have swelled over the years.

Difficult for users to understand and trust

People don't trust what they don't understand, and to illustrate that, I want to explain one of the ways companies spot patterns in data, which is via a technology called artificial neural networks, or ANNs. Such systems mimic the workings of the human brain, and literally learn over time which approaches work and which ones don't. They're used in personalization today to recommend books, music, and other products to customers and are just starting to be used to recommend documents to employees within a firm.

In a large company, with tens of thousands of employees, it's not difficult to envisage such systems being used to improve the firm's ability to make decisions about who should be promoted, receive special training, or be given special privileges, such as an expanded research budget.

To do their job, ANNs need large volumes of data, or inputs. In this situation, such data could include three years of promotion records, plus review records showing individual performance, and profit and loss figures showing how certain business units performed under the leadership of various managers.

Put all this information in the ANN, and it will start to tell you whether Bill should be promoted or Judy should be allowed to transfer to the Paris office. But there's a problem. It won't tell you why.

A trained neural network – that is, one that has received lots of data inputs, had plenty of time to practice making predictions, then learned from its errors – is a "black box". If you look inside, you can't see how it works. No one can explain why a particular ANN made a particular decision, not even the system itself.

But now you've created a situation in which your best judgement says yes, Bill deserves a promotion and Judy would thrive in Paris. It's just that ANN thinks both are horrible ideas, destined for failure. When you turn them down, your only explanation sounds like something a four-year-old child would utter. "Because. Just because."

Fear of the unknown is a good thing. It keeps us safe. Talk-radio hosts Ken and Daria Dolan, who dispense financial and life advice across the United States, are constantly saying, "Don't put your money in any investment unless you understand it thoroughly." The greater the importance of an action, the more critical it is that we understand its implications before we act. You don't need to understand how a bus works to spend £2 to ride across town, but you shouldn't buy the bus itself without understanding its current condition and future lifespan.

With very few exceptions, unless you can explain to users in a few sentences how a personalization function works, you shouldn't adopt it.

Difficult to maintain and support

Many of the current relationship-management software applications that enable one-to-one relationships with customers are driven by what are called business rules.

Business rules are the instructions that tell software, or people, how to operate. They are not software, but rather a reusable set of instructions that enables your organization to operate in a consistent yet flexible manner. Ronald Ross, co-founder of the business rules consulting firm Business Rule Solutions, llc, said, "Business rules are literally the *encoded knowledge* of your business operations."

"Upgrade platinum flyers to first class before gold flyers," is a business rule. So is, "Most valuable customers are those who order over £800 of merchandise in a year, or who place five or more orders a year that are over £100."

Business rules are not new; what's new is that many rules are being created and changed via interfaces accessible to non-technical business managers. Rules used to live deep within programming that was difficult and time-consuming to change. The resulting inflexibility of systems resulted in companies that are equally inflexible.

By "externalizing" some business rules, companies are seeking to be more adaptable and efficient, while they also get closer to their customers and deliver personalized service. But there's a problem.

The reality is that businesses now have these incredible new tools, but they have barely begun to build the skills necessary to exploit these tools. Few business managers have the training, expertise, or demeanour necessary to create and manage business rules, even when they are written in plain English. In early 2001, I wrote an article on business rules for *1to1 Quarterly*, the business journal published by Peppers and Rogers Group. Since so many CRM companies now rely on business rules to enable personalization, I wanted to find the schools, companies, and other institutions that are training business managers to create business rules. I couldn't find any. Zero. None.

Lest you think I was lazy in my research, I contacted numerous business-rules experts and all the leading business-rules software companies. Many of the latter provided general training regarding their software; none offered training that would help managers understand the complexity of delivering personalization to thousands or even millions of users.

As Bruce Schneier said, the operators' failure to understand a system makes the system vulnerable to security threats. This is partially because of what's called "social engineering", which is a fancy way of saying, trick people into doing something they shouldn't. One of the most famous hackers, Kevin Mitnick, claimed that 85% of his break-ins involved scamming people to give him information, and only 15% involved working on the computer. People who don't understand the systems they operate are much more likely to do and say things they shouldn't. This flaw doesn't only apply to security issues; it also applies to basic, everyday business activities. We've all experienced a simple example of such an error, when a newbie to a mailing list sends a message to the entire list, instead of to a single person. The result is that 5,000 people get the message, "How do I unsubscribe from this list?"

People are the weak link in technology, and you should never use a system that's too complex for your team to manage.

Too expensive

One excellent measure of complexity is cost. Complex systems are expensive. Probably the best defence against complexity is the simple reality that most businesses can't afford an overly complex system. But it's not a perfect defence.

Large consulting firms, systems integrators, and software firms succeed by selling large systems. They are very good at convincing executives that the only workable solution is one that will cost £5 million. In contrast, there's little economic incentive for anyone to invest time and money convincing you that you can accomplish a similar result through better training, or a subtle change in your process.

Not all expensive projects are flawed. Big companies are by definition complex, and it takes a lot of resources to serve millions of customers, or to pay hundreds of thousands of employees.

Whenever possible, you want to break up the risks you take into the smallest possible pieces, so that when you recognize a particular course of action is not going to pay off you can minimize your losses. If you do this right, you minimize your losses without sacrificing one ounce of the upside potential. Always keep your options open. Invest money now to preserve your option to keep investing resources at a later date. But, at the same time, always reserve the right to cut your losses at various stages.

Unlikely to produce the promised ROI

We never used to see stocks go up and down so quickly; industries used to change over decades, not a few years. People used to work at one company for an entire career. Wars were never carried out live on TV. Not long ago, a week was an acceptable time to mail a letter to a supplier and wait for them to respond with a quote. Overnight mail was just used for special occasions. Today, people get impatient if their email isn't answered within an hour, or less.

In every industry, in every size business, people are being confronted with tough decisions made even more challenging by increasing volatility. They have to decide whether they should invest time in building new skills or money in acquiring new capabilities and technologies. Many business leaders find themselves debating whether they should launch new Web-based divisions that would compete with their existing business.

New ways of competing are springing up faster than we can write or learn about them. I dare anyone to claim that they were able to predict that auctions would initially be the most successful business model on the Internet. Likewise, few thought that Amazon.com would either be the most successful brand online, or even that it would be able to survive as the number one online bookseller in the face of competition from the major established booksellers.

If you run a supermarket, you are probably thinking about whether you should be building an automated warehouse to be able to take Internet orders that you can deliver directly to people's homes. Webvan agreed to pay Bechtel Group over £500 million to build warehouses for it in 26 different cities. At the time, the company had only been delivering groceries to customers for five weeks, and only in the San Francisco Bay Area. As I write this in the spring of 2001, Nasdaq is considering delisting the company, because its stock has fallen below required standards.

That's unprecedented volatility, which makes it nearly impossible for anyone to predict the payback of a large, complex system. There are just too many variables: what will the competition do; will this technology be leapfrogged; will our people be able to manage it successfully; will our business partners cooperate; will the economy be healthy or anaemic, and so forth.

Lagging Changes

"OK, let me see if I have this straight," said Lawrence Cecil, president of Fantastic Toys Ltd, a £200 million a year entrepreneurial company that produces a wide range of entertainment and related products for teenage audiences. The three consultants across the table sat up a little straighter and got ready. They had been warned that Lawrence was the final hurdle before the project got the green light and that his relaxed manner hid a hard-nosed, highly analytical mind.

"We want to make our website more personal, and to have it serve as the hub of our relationships with the kids who love us. That means

tying together email, snail mail, telemarketing, and even the way we address our catalogues, right?"

The three nodded.

"You want to charge us £2.1 million to do that, and we've got £1.2 to spend. Just reverse these two little numbers in your proposal and we've got a deal."

The three started to smile, but each stifled the grin as they realized Lawrence wasn't smiling. "I wish we could do that, Lawrence," said Brian Roker, the senior partner. "But most of the costs in this proposal aren't even ours. The cost of the servers and software alone represents £900,000."

"Perfect!" said Lawrence. "Buy the software and servers, and we'll give you £200,000 to install them. Anything else we need to discuss?"

Brian looked at his colleagues.

"Lawrence was living up to his reputation. It wasn't even clear he had read the proposal. Lawrence, there are at least three stages of work. First, we need to assess the current state of your business and tailor the new software to fit your needs. Then we have to work with your teams to create new processes that take advantage of your new capabilities and that ensure everything works as designed. Finally, we need to train people to use the system, and we really should audit your current compensation systems to spot any disconnects between what people are being told to do and what you're paying them to do. That adds up to a lot of work. Truth is, £2.1 million barely covers it, but we'd really love to have bragging rights for working with Marvelous."

Lawrence sat back in his chair and lifted his hands, palms up. He swiveled left, then right, looking at his managers seated to either side of him. "Hasn't anyone explained our situation to these folks?

Heads nodded. One manager was about to start speaking, but thought better of it.

"Out of ten competitors in our niche 12 months ago, we're one of

two still standing. We survived because we do everything twice as fast and half as expensively as everyone else. We're not some bullshit big company that wastes more money in training programmes each year than we pay our entire staff. How much of your proposal covers the cost of stages two and three?"

The three hesitated.

"How much?" Lawrence demanded.

"About £800,000," said Brian.

"Here's the deal. I'll pay you £1.3 million to get the technology we need and install it. Or I can give the business to someone else. If you do it, my people will learn by looking over your shoulders, and they'll just have to read the manuals over the weekends. Do you want the work or not?"

"I see your point," said Brian, reaching his hand across the table. "We'll be thrilled to work with you on that basis."

The realities of today's ultra-competitive business world conspire to push training and process redesign far into the future. Hardly a business manager exists who doesn't appreciate the need for these steps, but when faced with hard decisions, most delay such expenditures as far into the future as possible. Managers perceive that process redesign, training, compensation reviews, brainstorming programmes, and any sort of change-management programme all slow a business in the short run, and no matter what the long-term benefits might be, judge that they can't afford to slow down. Given the increasingly personal nature of business transactions, however, this is a recipe for disaster.

Think about a company that acquires the capability to send email messages to hundreds of thousands of people quickly and inexpensively. Without robust new procedures to prevent mistakes, a well-meaning employee could accidentally or improperly spam thousands of people. In such a firm, two managers in different divisions could inadvertently send similar messages to the same people on the same day. An employee could send out a message that violates the laws in certain states or countries, because he was ignorant of the law and the firm lacks procedures to, at minimum, define the boundaries of proper use of the email system. As the power and reach of new technologies increase exponentially, so, too, must the care given to ensuring that there are reasonable checks and counterbalances.

A business I managed once fell into this trap in a highly embarrassing manner. We maintained an email newsletter and regularly received lists of names of people who wanted to be added to our mailing list. Typically, these came from department managers or association leaders who asked that we send the newsletter to their subordinates, or from one of our partners who came back from a conference with a pile of business cards individuals had given him or her specifically so they could start receiving the newsletter. When any of us had such a list, we would forward it to an internal person who typed names into the database. One day, a well-meaning employee of our firm found a list online of all the *New York Times* reporters who had made their email addresses public. The employee downloaded the list and sent it to our list manager, with a brief note saying something like, "Here's another group of people who should get the newsletter." I found out about this a few days later, when I received a letter from the *New York Times* demanding to know why we were spamming their organization. While the newspaper accepted our apology without humiliating us, this troubling event caused us to put a much more rigorous internal process in place to confirm that every submitted name was accompanied by an actual request from the recipient.

Most companies need outside help to ensure that someone thinks through carefully all the unintended consequences of new technologies and identifies the simplest possible measures to protect the company and its stakeholders from errors of judgement or execution. I say outside help, because the pace of business we're discussing generally precludes internal team members from taking the time – or having the imagination and thoroughness – to spot all the potential minefields on their own. Besides, just as frogs don't notice as the water around them warms towards the boiling point, it's often hard for people in an organization to recognize the danger that comes from seemingly subtle changes in their environment.

As I was writing the previous paragraph, a colleague walked into my office to ask my help in a privacy audit. The client is a large agricultural chemical company, which wants to know about issues for which it ought to be on the lookout. My colleague and I spoke about the time that has elapsed since most of the client's processes and procedures were created – probably 2–20 years – vs. the changes in business practices in just the past few years. In the past, the company never knew the names or identities of any end-users (translation: farmers), but now it was starting to collect, store, and use such information. Employees may be sharing personal information with their distributors or with retailers. The manufacturer's different business units may have different privacy policies, or none at all; are they reconciling differences or even spotting them?

The other side of this problem is that companies are spending huge sums of money on technology and not seeing the benefits they expected. This is not surprising, since technology alone is never a sufficient solution. If I had been present in the meeting with Lawrence Cecil, I would have recommended that he buy half as much technology, lower his short-term goals for personalization, and keep sufficient funds for training and new processes. But I wouldn't have used those words, because they are loaded with perceived negatives to someone like Lawrence. "Training" means higher costs, and "new processes" means bureaucratic bullshit. Instead, I'd talk about an all-out effort to lower costs and better leverage his existing assets. I might suggest lightning-fast interventions[4] with his different teams designed to produce immediate, measurable improvements in their results. The fact is that Lawrence is right: long training sessions in darkened rooms seldom produce lasting results. What does work are quick, practical – and most importantly – ongoing interventions.

Of course, despite my tactful wording, Lawrence would probably refuse to give us £600,000 for services of any sort, towards any suitably lofty goals. So I'd ask for a contract divided into quarterly or even monthly segments, which he could cancel at any point if our lightning interventions failed to produce measurable results. That provides us with a powerful motivation to not just do the right thing, but prove it, too. Delivering results will require ongoing, unending interventions – organizations have a tendency to slide backwards towards the way they've been doing things – but it also requires ongoing identification and removal of the obstacles that prevent or slow interventions from working. These could be outdated rules or compensation arrangements, political disputes between divisions, lack of expertise among employees, or confusion about a business's goals. In simple terms, you have to find out what isn't working, and fix it, quickly.

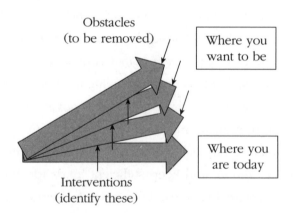

Figure 6.1   Fixing the Lag

Barry Stein, president of the consulting firm Goodmeasure, said that managers need less data and more thinking. Interviewed by *Computerworld*, Stein said, "Think, learn, examine. These are the tools to avoid unintended consequences." Stein described how 19th-century writer Charles Lamb imagined that humanity discovered cooking:

Hundreds of thousands of years ago, Lamb supposed, people lived in large extended families, with domestic animals, in crude houses built of wood and thatch. One day, a house caught fire; the only casualty was a neighborhood pig. When the residents returned, all that was left was a plume of smoke, a pile of ashes and a wonderful smell. Eventually, some of the people poked in the ashes and burned their fingers touching the carcass of the still hot, incinerated pig. When they put their burned fingers in their mouths to cool the burn, a delicious taste appeared. They had, Lamb said, discovered cooking.

Thereafter, when the people of the village wanted to celebrate, they picked out a house, put a pig inside it and burned the house down.

The moral, according to Stein: "If you don't understand what's cooking the pig, you are going to waste an awful lot of houses."[5]

One of the earliest unintended consequences that sprung up when companies started doing mass email campaigns to their customers was that customers emailed back. Companies received thousands of in-bound emails, and many companies had never even considered what to do if this happened. About three years ago, I led a two-day training programme for 30 executives from one of the leading personal-computer manufacturers and told the executives the first morning that the previous day I sent an email to the "get information" email address found on their website. At that point, about 24 hours had elapsed with no response. The highest ranking executive took special note of this and made me check my email at every break. We got through the entire two days without a response, and the whole time everyone was expressing surprise and growing outrage at this supposedly atypical event. Turns out it wasn't atypical; the company simply wasn't prepared to handle the volume of inbound emails it received.

Watch Out

Nobody has a crystal ball, and I'm certain that we will all be surprised by unintended consequences that seem to materialize out of thin air and suddenly fill global news headlines. That's what happens when you mix technology and business. So while I'm virtually certain that this list is incomplete, what follows is a list of the unintended consequences for which companies should be watching. To avoid these consequences, compare each new use of technology and each change of an existing process against this list, and be sure that safeguards are put in place to prevent these or other negative outcomes from becoming real.

Head-of-state scrutiny

Nobody's perfect, and few of us can stand up to the type of scrutiny that goes with the territory of being a president or prime minister. Yet, soon we all could be under that type of scrutiny, simply because every detail of our lives will be tracked. The everyday minor elements of our lives that have largely gone unnoticed will increasingly be recorded and analysed, and ultimately be accessible to others. As companies knock down the walls between databases and departments, they will be able to amass a near-complete picture of an employee's or customer's activities. Today, the records of your mobile-phone calls exist in a different database or department from those of your office calls or your email activity, but the push to build a complete view of customers – and eventually employees – will erase these disconnects.

We can stray slightly over speed limits, take occasional long lunches, say we're working from home when we're actually shopping for a present, and take hundreds of other small actions that are in the grey area between acceptable and unacceptable behaviour. It's not worth anyone's time to find out if we violate perfect behaviour, because doing so takes effort.

But in the not-too-distant future, not only won't it be difficult to spot such violations, but it will also be near-impossible to miss it ... unless companies recognize that not everything that can be tracked should be. Grey areas are good things, providing each of us with the breathing room to be less than perfect, which is a basic condition of human life.

If we don't limit the scrutiny of individuals, there will be only two other options, neither of which is particularly appealing. We could make overt decisions to allow rules to be routinely, and obviously, violated, or we could begin to enforce such rules with much greater vigour than is now the case. The first option fosters disrespect for laws and the rules that bring us order and leaves too much room for rules to be enforced selectively or in a discriminatory fashion. The second option could force businesses to treat people in a manner that is counterproductive, just to avoid the appearance of discrimination or unfairness. No one questions how long you take for lunch if you are the best performer in your department. But when there are digital records available of all the people in your department and how long they take for lunch each day – and yours is twice is long as everybody else's – it may become difficult or impossible for the company to allow that behaviour to continue.

Over-aggressive employees

The vast majority of companies today motivate their employees to meet the company's short-term financial goals, not to strengthen long-term relationships with key stakeholders. Employees are compensated to generate and conclude transactions that bring the company growth and profits. This is natural, healthy, and a consequence of having a free-market system. But individual employees who are motivated and highly compensated in this manner will not be sensitive to the privacy concerns of other employees, customers, or suppliers. "In a perfect world," they might say, "we'd get permission before using personal information. But I don't get paid to get permission. I get paid to generate profits."

A complicating factor is that most employees don't understand the consequences of their actions. Of the dozens of firms who have been vilified in headlines recently for invading privacy, not one employee has admitted that they deliberately invaded privacy. Most seem to have been taken unaware that their practices would have such negative – and highly public – consequences. Unfortunately, this doesn't provide any comfort to the people whose privacy gets trampled.

Aiding criminal behaviour

As companies start to amass increasing amounts of personal information and as they make personalized interfaces available to more of their stake-holders, they are increasingly exposed to the consequences of criminal use of this data. Just because a company takes safeguards to ensure, for example, that data is password protected, or if it uses a biometric identifier to protect security, it doesn't mean that someone can't misidentify themselves and steal information from the system.

If that is the case, centralized access to personal information makes it far easier for someone with criminal intent to engage in stalking behaviour or identity theft, to steal assets from a person, or to find out details of one's personal life that the person never intended to share with a dangerous criminal.

There are many facts that a person might not want such a criminal to know. It's not hard to envisage a new breed of subversive data analysts, who sort through corporate databases looking for enough negative infor-mation to allow them to blackmail a person. Some could even learn to terrorize individuals by bluffing that they know more than they actually do. If someone gained access to my bank accounts and stock positions by stealing information from my bank, they would be able to give me the impression that they could steal all my money. This could give them tremendous leverage over me.

Mistakes that never die

The more data firms collect, the greater the likelihood there will be mistakes in that data. The more widely data is exchanged between companies, the greater the consequence of the mistakes. Companies have a much greater ability to proliferate data than individuals have the ability to correct mistakes. There is also, as we talked about before, a tremendous tendency that people have to believe data once it gets into a company database, because many of us trust authority figures and vest authority in large companies and seemingly elaborate national or international systems.

If a company has less than perfect ability to keep its data correct – and most companies do – then it runs the risk of sending ripples through thousands of people's lives that could cause those people untold hours of problems. They will blame these problems on the company that originated the mistake.

Companies need to make it easier for people to find and fix mistakes, but more importantly, they need to be much more cautious in sharing data with other firms. A deadly virus doesn't pose much threat when it's contained in a secure medical laboratory, but when it infects a person who then travels on seven planes across three countries, it becomes a global danger. In the same manner, mistakes in a closed system – a single business unit or corporation – are relatively easy to fix. But as soon as a firm shares data with other firms, who have the option of doing the same, mistakes can spread like viruses.

Revealing secrets unintentionally

Several times I've been at a party and met the spouse of someone with whom I work. Somewhere in our conversation, I mention some seemingly innocuous detail about work – what a great job Sue did at last year's conference in Orlando, or how lucky we were that Peter was able to step up when we were in danger of losing our biggest account – only to see a look in that person's eyes that tells me I revealed some piece of information that somehow struck a nerve. Sometimes this may be the result of small things, such as Peter's wife being upset that he doesn't talk more readily about his successes at work. But what if, somewhere in my description of Sue's performance in Orlando, I inadvertently revealed that she lied to her husband, perhaps to cover up an affair?

The more data companies collect and share, the greater the odds that secrets will be revealed unintentionally. In nearly every murder mystery, movie about adultery, and news report about captured criminals, it's some small detail that reveals the perpetrator. The detail alone means nothing – say, the presence of a matchbook from a certain restaurant – but when combined with other details, it reveals a secret. Not all secrets hide illegal or immoral acts. We keep secrets to avoid hurting someone's feelings, to avoid getting a person excited before we know for certain whether a positive outcome will come to pass, and to protect our own desire for privacy and personal space.

Violating laws in other jurisdictions

National and local governments exhibit widely varying degrees of concerns about privacy, and some may act without regard to the practicality of corporations' attempting to comply with overlapping, perhaps contradictory laws. The European Data Directive, for example, forbids the export of data to countries without adequate privacy protection. Such a prohibition could stop a European subsidiary from transferring information to its American parent company, it could make it difficult or impossible for American firms to sell their products directly to European consumers, and it could stop American firms from participating in, or accessing the results from, European market-research programmes. This is because the European Commission has not ruled that American firms provide such protection. In an attempt to bridge this gap, in July 2000 the US Department of Commerce adopted a set of "safe harbour" privacy principles, which if followed voluntarily by a US company, would enable that company to meet the standard of adequate protection. However, as of early 2001, only 12 US companies had applied for safe-harbour status.

The evolving nature of privacy laws is one of the primary factors motivating companies to appoint chief privacy officers, or CPOs. A recent PriceWaterhouseCoopers study listed "tracking pending legislation" as one of the most common and important CPO roles and revealed that 47% of companies place their CPOs in either the legal department or the office of general counsel.[6]

But it's hard enough for the CPO to keep track of all the laws that impact a multinational company, never mind to convey this information to the multitude of employees who today manage personal information. To further complicate this situation, individual states and provinces will likely step up their attempts to legislate privacy. In the United States, for example, consumer protection has primarily been the domain of state law. States such as New York, California, Michigan, and Florida often act before the federal government and often begin movements that result in eventual changes to federal law.[7]

Immortal Memories

8 July 2002

It's well past midnight, at the end of one of the most aggravating weeks you've had in years. You had a fight with your boss, your spouse, and two of your three kids. You're surfing the Web, trying to calm down before you go to sleep, and you discover a discussion forum at an industry site. Shaking your head at the comments of some so-called expert, you dash off a lengthy response, coloured with frank language and what could be interpreted as a personal attack on the credibility of the original author. By the next morning, everyone is once again happy, and life looks much better.

12 December 2008

You get rejected for a job because a routine background search discovers a pattern of "abusive public postings" stretching back to 2002.

Partner missteps

The best privacy policies and the most disciplined processes within a company won't stop misuse of individual data if a company shares it with its partners and doesn't have the same level of control and discipline over the ways that the partner uses individual data. It is one thing to say, "We will impose the same requirements on our partners as we do on ourselves," but it is enormously difficult to do it. Each company has different motivations, goals, cultures, and ways of profiting. Few have enough clout to dictate how the partners operate internally.

The simple truth is the more a firm shares data with other companies, the greater the likelihood that one or more of those companies will do something that will harm people as well as the company that shared its data. Every caution listed in this book is a potential trap for a firm's partners. The more partners, the greater the odds that one gets caught.

With so much that can go wrong, how can companies avoid certain disasters springing from the use of personal data? It's not hopeless. Firms simply need to set boundaries of acceptable behaviour around the processes and technologies they utilize. These boundaries not only protect privacy, but they also increase the odds that personalization will generate the growth and profits that companies need to survive. It's time to look at a few simple rules that can help companies create such boundaries and put them in the right place.

Thought Exercise ...

What Consequences?

Pick three initiatives under way in your department or company. To make life more interesting, add a few others that aren't under way but that you or others have entertained ... Look for ideas that stretch the envelope, broadening either the services you offer or even the definition of your business. Now, take these six initiatives

Initiative:
_____

Negative:	**Example:**
Head-of-state scrutiny	_____
Over-aggressive employees	_____
Aiding criminal behaviour	_____
Mistakes that never die	_____
Reveal secrets unintentionally	_____
Violation of laws	_____
Immortal memories	_____
Partner missteps	_____
_____	_____
_____	_____
_____	_____

Positive:	
_____	_____
_____	_____
_____	_____

Figure 6.2   Unintended consequences

one at a time and try to think of one example for each of the negative consequences listed here. Since your business is unique, you'll probably be able to come up with a few other categories of consequences, so add them to the bottom.

Not all unintended consequences are negative. Few inventors anticipate the most compelling uses or benefits of their inventions. List a few examples of positive outcomes that others don't see and that may be too speculative to serve as the primary reason to pursue this initiative.

Notes

1   From Rocky Mountain Institute website
    http://rmi.org/sitepages/pid157.asp
    but also repeated nearly word for word at many other sites.

2   "The Informational Value of Eyewitness Responses to Line-ups: Exonerating Versus Incriminating Evidence," by Gary L. Wells and Elizabeth A. Olson, Iowa State University
    http://eyewitnessconsortium.utep.edu/wells%20and%20olson,2000.pdf

3   *Secrets & Lies: Digital Security in a Networked World,* by Bruce Schneier, 432 pages, 1st edition (14 August 2000), John Wiley & Sons Inc., New York; ISBN: 0471253111.

4   Thanks to Ron Cox, former president of North American Operations for AchieveGlobal, who is now a consultant working with a number of companies to develop and implement strategies for growth and change. He provided the "Fixing the Lag" chart as well as my comments in Chapter 7 about positive and negative immediate consequences.

5   "IS and the Art of Cooking Pigs," a sidebar published in *Computerworld's* January Leadership Series, *Be Careful What You Wish For: Managing Technology's Unintended Consequences,* by Vaughan Merlyn and Sheila Smith, 20 January 1997
    http://198.112.59.30/home/online9697.nsf/all/970120leadership

6   "The Chief Privacy Officer: Putting a Public Face on Privacy Management,"
    by Ruth Nelson, January 2001
    http://www.pwcglobal.com/extweb/newcolth.nsf/docid/
    732AC589FCF6A523852569BD005A07CB?OpenDocument

7   "Privacy Legislation in the States," *Privacy and American Business*
    http://www.pandab.org/v5n3priv.html

7

Setting Boundaries

F air warning: the next few pages – but hopefully not the entire chapter – show why people fall asleep when business meetings turn to the subject of privacy. It's the only place in the book where I couldn't avoid using "subparagraph a" and similar references. But even if you skip the first few pages, please don't leap past the whole chapter, because this is where I share with you how a few changes can make a dramatic impact on your business. These changes go over and above the type of issues that most experts discuss when companies focus on privacy.

The majority of privacy laws around the world are based on the principles of fair information practices (FIP), which were published in 1980 by the Organization for Economic Co-operation and Development, an organization of 30 countries committed to "a market economy and a pluralistic democracy." The principles were derived from work done by Alan Westin, who has been professor of public law and government at Columbia University since 1959. In his 1967 book, *Privacy and Freedom*, Westin said that the right to privacy is the "claim of individuals, groups and institutions to determine for themselves when, how and to what extent information about them is communicated to others."[1] I agree with this definition and believe that companies should adhere to these principles, but they don't provide the whole answer. FIP doesn't show companies how to profit from personalization while at the same time protecting individual privacy.

The central message of this book is that personalization, delivered properly, protects privacy. FIP principles are important and provide a useful guide for companies trying to respect individual privacy as well as the rights of other institutions. But they don't tell companies *how* to motivate employees to simultaneously strengthen relationships and respect privacy. FIP doesn't tell us what's in it for the company. Too often privacy advocates, consultants, and corporate executives ignore the realities of how businesses work. The way they talk about privacy creates a perception among company employees that respecting privacy is similar to paying taxes: it's no fun, but if you don't follow the rules you'll get arrested. As a result, privacy has become the province of lawyers and detail-oriented administrators. It's viewed as a boring necessity, rather than an area rich with opportunities.

After this brief tour of fair information practices, I'll offer an additional set of guidelines designed to help companies earn the loyalty of key stakeholders in a manner that respects both their wishes and their privacy. Here are the general principles adopted by the OECD in 1980:

Collection limitation principle   There should be limits to the collection of personal data and any such data should be obtained by lawful and fair means and, where appropriate, with the knowledge or consent of the data subject.

Data-quality principle   Personal data should be relevant to the purposes for which they are to be used, and, to the extent necessary for those purposes, should be accurate, complete, and kept up to date.

Purpose-specification principle   The purposes for which personal data are collected should be specified not later than at the time of data collection and the subsequent use limited to the fulfilment of those purposes or such others as are not incompatible with those purposes and as are specified on each occasion of change of purpose.

Use-limitation principle   Personal data should not be disclosed, made available, or otherwise used for purposes other than those specified in accordance with the purpose-specification principle except:

(a) with the consent of the data subject; or

(b) by the authority of law.

Security-safeguards principle   Personal data should be protected by reasonable security safeguards against such risks as loss or unauthorized access, destruction, use, modification, or disclosure of data.

Openness principle   There should be a general policy of openness about developments, practices, and policies with respect to personal data. Means should be readily available of establishing the existence and nature of personal data and the main purposes of their use, as well as the identity and usual residence of the data controller.

Individual-participation principle   An individual should have the right:

(a) to obtain from a data controller, or otherwise, confirmation of whether or not the data controller has data relating to him;

(b) to have communicated to him data relating to him within a reasonable time; at a charge, if any, that is not excessive; in a reasonable manner; and in a form that is readily intelligible to him;

(c) to be given reasons if a request made under subparagraphs (a) and (b) is denied, and to be able to challenge such denial; and

(d) to challenge data relating to him and, if the challenge is successful, to have the data erased, rectified, completed, or amended.

Accountability principle   A data controller should be accountable for complying with measures that give effect to the principles stated above.[2]

Intellectually, these principles make sense. But they represent best-case standards and adhering to them means an uphill battle against everything else that's happening in a growing business. Take the security-safeguards principle, for example. No business executive would dare argue against this principle, yet, as we've seen, data security at many companies is dubious at best, and growing more precarious each day. The purpose-specification principle says that a company should tell individuals up front how it will use a particular piece of information and then limit use to that specific purpose. But this runs counter to the interests of most companies, where marketing, sales, and other groups are paid to maximize the value of all their assets, including information assets.

Don't get me wrong; I believe in these principles. The problem is that they won't be practised by large numbers of companies unless companies make significant changes in the way that they measure and reward the performance of employees and business units. Fortunately, it only takes a few changes to create an environment that fosters the right kind of relationship and the right respect for individual privacy. The philosophy underlying these changes views privacy as the natural by-product of healthy relationships between companies and individuals. Not many such relationships exist today, because companies are managed in a manner that often creates adversarial rather than collaborative relationships between a company and its customers and employees.

Two Critical Changes

By making two changes in the ways employees are compensated, any company can simultaneously become more profitable and achieve the right balance between privacy and personalization.

Change #1: Compensate employees to satisfy more needs of existing customers

In most corporate cultures, employees do what they get paid to do, not what they are told to do. To achieve real change, firms have to redesign their compensation systems to reward desired behaviours and punish negative ones. If employees are compensated to consistently satisfy more needs of each existing customer, they will naturally seek ways to use personalization to each customer's benefit. That's because personalization capabilities are all around us, and they represent a potent tool that can benefit all the parties in a business relationship. Using these capabilities in a manner that upsets a customer would drive that customer away, and employees won't do this if such an outcome harms their own welfare.

Readers who are familiar with the principles of one-to-one marketing and/or Customer Relationship Management may be surprised that I haven't suggested that the first step is to identify which customers are most valuable to your company. This is an important step, but it's not as important as this broader change I am proposing. Here's why.

If you compensate your employees to satisfy the needs of existing customers, they will naturally gravitate to the customers with the financial means and the depth of needs to warrant extra attention. Likewise, they will minimize the time they spend on customers who care only about price, who are never happy, and who demand their money back no matter how hard you work to satisfy them.

In other words, if you make the two changes I am suggesting, all the other steps necessary to create more profitable customer relationships will happen as an inevitable result. By the same token, your organization will become more sensitive to the needs of other stakeholders. The habits of listening and responding are catching.

In my experience, most privacy abuses stem from efforts by companies to use personal information to *acquire* new customers, not to better serve existing customers. I'm constantly frustrated, sometimes to the point of amusement, that 90% of the discussion at personalization and customer-relationship management conferences revolves around tools and tactics for acquiring new customers. Despite the fact that many of the participants are leading global companies, a fly on the wall would conclude that few of the attendees actually have any customers yet.

Most privacy abuses occur because employees get compensated for generating revenues, not for growing relationships. What I'm proposing is that they must do both, and that the simplest way to achieve this is to compensate employees for their success in keeping existing customers happy. Such a policy has a secondary effect of motivating employees to only acquire new customers likely to be most happy with the company's services. Here, too, this lessens the scope and likelihood of privacy abuses; employees will have nothing to gain from chasing prospects who don't value the company's offerings.

Am I suggesting that companies stop paying commission on sales to new customers? Whenever possible, yes. For most companies, it's far more profitable to compensate salespeople, and others, for subsequent – rather than the first – transactions. Employees need to succeed three times to conclude two transactions:

1. they must find the right kind of customer;
2. they must immediately satisfy the new customer; and
3. they must immediately seek to satisfy more of the customer's needs.

When companies reward employees simply on the basis of revenue, they foster all sorts of potentially negative outcomes. It doesn't matter whether a customer is happy with her purchase, just as long as she doesn't ask for her money back. Employees under pressure to meet quotas can recruit the wrong kind of customers who will cost a company far more in service and support costs than they contribute in revenue. The end result is a "show me the money" sort of attitude that encourages employees to look at customers and other stakeholders as targets rather than people with whom the company can build win–win relationships. This sort of attitude is extremely dangerous as the data trails grow behind each one of us.

Focusing on the strength of relationships with existing customers will create a culture in a company that values lasting relationships, and the benefits of such cultures will lead to stronger relationships with other stakeholders as well. Just as an internal culture of personalization is necessary to foster deeper relationships with customers, so too is the right external focus necessary to enable attitudes that benefit employees and other stakeholders.

Change #2: Develop modular capabilities

To make the first change, companies need the flexibility to accommodate the differences between individuals. The best way to accomplish this is to motivate employees, suppliers, and partners to develop modular capabilities that can easily adapt to changing needs. This is the practice of mass customization, which is basically a Lego building blocks approach to business. Every time that stakeholders make a decision about developing a new capability, they should maximize the degree to which that capability can be flexible. In the late 1980s, when PCs were just starting to penetrate households, many people considered buying word processors instead. Their early success stemmed from the fact that people couldn't conceive what else they would do with a PC besides write. Eventually, people recognized that you can do just about anything with a PC and that it's foolhardy to spend nearly the same amount of money to acquire a piece of equipment with just one capability, instead of an unlimited number.

This isn't just a "nice to have", and thus the development of modular capabilities ought to be another critical measure when determining employee compensation and when bargaining with suppliers and partners. The more flexible a company's supply chain, the more flexible the company will be, and hence its ability to accommodate individual needs.

As we discussed in Chapter 6, modularity brings with it potential problems in terms of both security and excessive complexity, but there is no escaping the need for modularity to fulfil both corporate and individual needs. Be watchful for these potential problems, but don't avoid modularity because of them. Without modular capabilities, no firm will be able to succeed in the years ahead.

Implemented together, these two principles create a corporate culture in which everyone is compensated to constantly listen to customer feedback and then do something better for the customer as a result. Personalized relationships will emerge naturally, because all the elements are present that foster such personalization: the company is rewarding win–win relationships and developing the modular capabilities necessary to support them. At the same time, there's no economic incentive for employees to intrude on people who wish to be left alone.

Modular capabilities make it profitable for a firm to support personalized relationships. Customization becomes routine and cost-efficient, and in many cases costs will go down, not up, as firms start to tailor the way they treat each valuable stakeholder. Much of this savings comes from the elimination of waste and the reduction of inventory levels. Often, it is cheaper to customize a product in response to a new order than to hold inventory in stock waiting for an order to come in.

Firms that fully embrace these two principles will adapt faster to changing markets. In a traditional product-driven firm, most feedback is ignored. Customers tell your salesperson their needs, and that salesperson tries to sell one of the three products she has to offer. There's little she can do with such feedback, because the firm has a limited and inflexible product offering. But a firm that makes these two changes is able to respond to feedback by customizing the products and services it delivers.

Almost every time a stakeholder says, "I wish this was different," you can make it different.

In such a firm, change and adaptation are natural; they happen daily, one person at a time. Contrast this with product-driven firms, where change can be wrenching because it happens so rarely.

By focusing employees on these two changes – developing modular capabilities and constantly broadening share of customer – firms will motivate them to make the right decisions. They will evaluate vendors and technologies based on their ability to support modularity. They will seek out customers, partners, and employees who are most likely to respond well to a constantly expanding set of services. They will reject schemes that have any possibility of upsetting or scaring off valuable customers, since such defections would cause them significant financial harm.

Making It Happen

With these two changes at the core of an organization, we can now see the big picture of how companies can balance personalization, privacy, and profits. To make this happen, we simply need consequences that reinforce the changes we seek, plus four other principles to help manage the practical aspects of using technology to strengthen business relationships.

Making change stick: PIC-NIC

Personalization is about changing the behaviour of people. It's about serving people in such a way that both they and the company profit. To do this, we need to understand a reality that behavioural psychologists and managers in many fields have recognized for some time. Ron Cox, former president of AchieveGlobal and currently an independent management

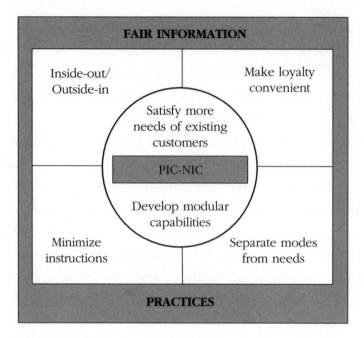

Figure 7.1   Balancing personalization, privacy, and profits

consultant, says people need to have either positive immediate consequences (PICs) or negative immediate consequences (NICs) to their
actions. "The threat of a potential problem somewhere down the road
doesn't change behaviour," Cox claims. "The fact that someone may sue
you someday isn't enough to get most people's attention. The threat of
something down the road doesn't work."

What works, says Ron, is you do something, and something else
happens, right away.

I once participated in a workshop that demonstrated this principle to
great effect. The workshop leader sent three people out of the room, telling
them that, when they returned, they would have to find an object in the
room that the rest of us had selected. The room was huge, and the
searchers received no other hints or instructions.

The leader then asked the rest of us to try various techniques to guide
each searcher's quest. In the first case, we were instructed to do nothing.

We remained completely silent, and the searcher became extremely un-comfortable and spent many embarrassing moments wandering the hall in vain, never finding the object. For the second searcher, we were instructed to boo and hiss loudly anytime the person moved away from the object. This person found the object eventually, but later admitted that doing so was a hideous experience. For the third searcher, we were told to boo and hiss when they moved the wrong way, but to also clap and cheer when they moved in the proper direction. This person found the object quickly and enjoyed the process.

You might want to try this demonstration at your next group meeting. It demonstrates vividly that people need both positive and negative conse-quences and that the lack of such consequences – or the presence of only one – makes doing our jobs much more difficult.

Shifting from a mass production to a personalization culture necessi-tates major changes in behaviour and attitudes among employees. The only way such changes will occur is if they are supported by immediate positive and negative consequences. Although I've proposed here a limited set of changes capable of producing such a shift, they don't have a prayer of taking hold – or of sticking – unless backed by tangible consequences. Few senior management teams are willing to change reward systems to reinforce such changes, but this hesitancy keeps companies in a dangerous place in which they both fail to earn increased loyalty from valuable stakeholders while continuing to threaten individual privacy. Middle and lower level managers seldom have the authority to change measurement systems on their own, and they need to recognize this fact and escalate quickly this issue to the appropriate levels.

For example, if a salesperson doesn't capture in a customer database important information about customers, such as what he or she has learned about them, their needs, and the manner in which they make decisions, then the company should not pay the commission. There has to be an immediate consequence, and not getting paid until you do something is a noteworthy consequence. Of course, these consequences have to be reasonable. No amount of reward or punishment will enable people to do something they are not capable of doing.

The same principles apply to changing behaviour patterns among stakeholders. When a customer does something that the company considers to be the behaviour they wish to foster, there should be an immediate positive consequence. The customer should get better service, save money, and/or have an important need satisfied. When firms are trying to implement personalization, employees need to have either positive or negative immediate consequences based on their ability to support the new practices that the company is trying to foster.

Inside-out/Outside-in

As you know, I advocate an Inside-out/Outside-in approach to enabling individuals to control how companies use their personal information. A company doesn't have to adopt this approach to make the following guidelines work, but I urge you to consider doing so. When a company intends to collect information and share it with other companies or with individuals outside of the firm, this should only be done when individuals opt in to such practices, which means that they approve it in advance. This is not the way that most companies do business today. Most companies have an opt-out policy, which puts the burden of action on the individual. But the potential consequences of sharing data in an open system network are too significant to put the burden of action on the individual. If a company wishes to profit by sharing personal information with other companies, then it should be willing to take on the burden of securing permission from an individual before it shares information in this manner.

However, if a company chooses to use personal information only within its own organization, then I think we need to acknowledge the reality of business practices today and say that, in this circumstance, the current practice of opt-out is sufficient. It's not ideal; individuals still will have to raise their hand to prohibit a practice to which they object. But in practical terms, there are tremendous challenges to dictating that companies can't use information even within their own walls without first obtaining permission. Individuals interact with companies every day and companies need access to this information to complete their basic functions and to complete the tasks that benefit their customers, employees, and investors. Companies have a right to remember what they do and what they have done in the past, just as individuals do. The key, however, is to blend this approach in with the guidelines that follow, so as to create a pattern of behaviour that will be conducive to individuals, as well as companies.

Minimize instructions

The only reason that personalized business relationships are now possible on a mass scale is because companies are using technology to manage interactions with individuals. To make this happen, companies need to provide instructions that govern how each technology supports these inter-actions. The cardinal guideline for success in this environment is to use the absolute minimum number of instructions possible. Use too many, and it becomes impossible to manage them. Instructions conflict with each other, cancel each other out, and hamstring your systems so that neither technology nor people work effectively. This guideline applies to all sorts of instructions, from tactical ones that dictate how systems work to enterprise-wide ones that provide the glue that binds different units together. For example, a firm that wishes to give freedom to each operating unit, while still making it possible for units to share data about customers, might create a rule that says every customer database must use the same format as a customer-identification code. The units will have the freedom to organize their databases any way they wish, and to use whatever software they choose, just as long as this one field exists. In this way, the company can recognize the fact that the same customer appears in multiple databases.

These sorts of instruction used to be created and managed exclusively by programmers, but there are a number of problems with having programmers totally responsible for the manner in which software operates. For one thing, there aren't enough talented programmers, and businesses have to wait far too long to get programmers to make every change. Companies can't be nearly fast or flexible enough. Another problem is that programmers know a lot about programming, but they don't necessarily know that much about a business or the needs of the firm's stakeholders. So while they might choose a suitably efficient or technically elegant solution to a programming challenge, they may not recognize a course of action that brings a business success.

For these reasons and more, with increasing frequency the people who use software – business managers and other employees – are gaining the ability to tell the software how to operate. In most cases, the way they do this is through business rules.

"Almost every one of our clients is pursuing a strategy of having business managers have control over the basic rules," said Alan Crother, an executive at Adjoined Technologies. "They want to allow business units to specify and make changes to rules quickly, without heavy IT involvement or coding."

Think about the best employee you have ever managed. Wouldn't you like to be able to capture what makes her so talented and imbue those qualities into an automated system that could deliver extraordinary results time and again? That's one promise of business rules, but the challenge is translating her approach into a manageable number of repeatable principles. You could, for example, use the example of a wonderful office receptionist to inspire rules regarding how a website should greet and assist new visitors:

- be friendly but businesslike as you ask the reason for the person's visit;
- immediately contact any people (or databases) who need to know the person is visiting;
- tell the person what to expect in terms of waiting time and how they will be treated;
- if the person needs to wait, make them as comfortable as possible.

Ken Molay, director of product marketing at Blaze Software (now part of Brokat), explained that business rules can also be used to balance the needs of a customer with those of a company. "The goal of personalization isn't just to give the customer whatever they need," he said, "but rather to keep them happy while also ensuring that the company profits from delivering the service."

It's not hard to create a simple business rule, such as "employees must have a job-performance rating of 'satisfactory' or better in order to be eligible for a year-end bonus." It is difficult, however, to create and manage a full complement of rules, especially when the goal is to personalize the way a firm treats individuals.

As rules multiply, the odds increase that they will cancel each other out or will cause unexpected results. This is especially problematic when numerous departments interact with the same employees, or multiple business units target the same customers. What if two different business units have different privacy policies, and both collect information from the same customer? Which policy takes precedence?

So here's the situation: with increasing frequency, employees who have little experience with business rules will be creating them, and these rules will influence greatly whether a company succeeds or fails in its efforts to balance personalization, privacy, and profits. That's why I say, the fewer the rules, the better. It's highly productive to invest extra time in crafting the smallest possible number of instructions. But there's another major benefit to this tactic. The more restrained a firm is in creating rules, the more freedom employees have in finding ideal solutions for customers and other stakeholders. Think about a customer-service operative dealing with a disgruntled customer. If the employer specifies exactly what that operative can or can't do to keep the customer happy, it is likely to foster a culture in which operatives don't even try to help customers, because they run out of the time and energy to absorb, never mind follow, the rules. In contrast, a firm could create a single rule that says the operative may take any action necessary to keep the customer happy as long as the costs of that option represents no more than 20% of the profit the firm derived from the company last year. To accomplish this, the firm needs to be able to give the operative real-time access to what that profit number is. If that's too difficult a challenge, a company could use a simpler number such as the revenue generated during the last year. The point is that you are setting an understandable and reasonable boundary. You are saying you can go up to this line, but no further, and within this range you have the freedom to find the best solution.

Make loyalty convenient

The core benefit of personalization is making it more convenient for a person to do business with a company, but most companies invest far too little time in mapping strategies and services that accomplish this.

This principle should be used as a test to determine whether a proposed service or strategy offers the correct balance between personalization and privacy. Business units should only pursue initiatives that make loyalty more convenient for valuable stakeholders while at the same time providing a suitable benefit to the business, such as a desired return on investment. It's startling the number of programmes created by companies that do absolutely nothing to make loyalty convenient. Introducing or promoting a new product, for example, doesn't make loyalty convenient unless that product is capable of remembering knowledge about a customer gathered by an earlier product. In far too many situations, it is just as easy for a customer, investor, or employer to do business with a competitor than with the current company.

Separate modes from needs

Personalization succeeds when companies learn to accommodate the unique needs of individuals. But not every service a company provides revolves around unique needs. As much as people have numerous differences, they also have many things in common. I use "modes" to describe a small number of common activities in which we all engage at various times: buying a gift, bargain hunting, browsing, learning, and hurrying. It makes sense to build functionality that supports each mode into the areas where individuals interact with businesses; doing so saves people a lot of time and increases the odds they will buy from the company. Fortunately, customers will quickly self-select a mode, if you give them the option.

What separates needs from modes? In the case of personalization, a need is persistent; it doesn't come and go. If you have eight kids, you are always buying in bulk. If you run a technology company, you are always interested in recruiting talented engineers. But modes change constantly; Thursday you might be in a hurry, while Saturday you may be eager to explore.

Imagine a mall with guides who tracked visitors the way some personalized websites do. They would try to deduce your intentions by noting every time you paused to look in a display window, turned around to get your bearings, or responded in a friendly manner to a clerk's greeting. The result would be laughable, probably resembling a *Monty Python* sketch:

> **Mall guide**   *You said you were interested in satin doilies.*
> **Customer**   *No, I didn't.*

> **Mall guide**   *Yes, you did. Your eye lingered on them for eight seconds.*
>
> **Customer**   *I was looking at the bug crawling across your display.*

While it's possible to learn about a person's intentions in as few as three clicks, in most cases you are not learning about that person. You are learning about what "mode" they are in. Think of it this way: businesses should offer modes to make life easier for all their customers, but then offer customers the option to personalize their modes, thus making loyalty even more convenient. An example: let people save information about others for whom they buy gifts on a regular basis.

When we separate clearly modes from needs, we strip away most of the excess data that makes it hard to understand what each person needs. Modes change each time a firm interacts with a person and can be deduced quickly. Needs – especially the most powerful needs – emerge over a longer period of time. By recognizing modes, firms can provide meaningful customization before they know a single thing about a person. As the customers come to appreciate such customized service, they become much more likely to share information about their needs. But even if they don't, by understanding and accommodating modes, companies can deliver many of the benefits of personalization without having to first collect significant amounts of personal information.

You may be thinking: aren't modes simply functional capabilities, such as checking an order or performing a search? No, modes are more complicated than that. "In a rush" is a mode, and so is "bargain hunting". Both are common mindsets that could lead a person to appreciate a wide variety of functional capabilities and that require more than a single capability to serve. To illustrate this, think of a distributor who needs at various times to access details about a manufacturer's products, services, prices, and lead times. Two modes could be "with a client" or "not with a client". In the former situation, the distributor will be focused on the client's specific needs and will not be interested in accessing any general information about the distributor's dealings with the manufacturer (such as total orders outstanding or available credit). Instead, the distributor will want the ability to describe the client's needs and specifications and quickly access services that provide the client with what is needed. In the latter situation, a distributor will probably want access to all information, since it needs to manage both client orders and its own status with regards to the manufacturer.

By organizing functional capabilities around certain modes, companies reduce dramatically the need to collect, store, and analyse individual data. This reduces processing time and infrastructure needs, minimizes privacy issues, and enables a company to be more selective about those times when it must request information from a person.

Thought Exercise . . .

Invent New Services

Pick a stakeholder group and develop four to six modes that summarize the most important set of activities in which those stakeholders are likely to engage. Then list as many services as possible that your firm could provide to support each mode. These services should be a mix of capabilities that you already provide and those that you could develop to further strengthen these important relationships. The primary advantage of gathering all these services together is that you make it far more convenient for stakeholders to do business with you, once it becomes clear to either of you which mode best addresses their current mindset.

Whether an action then takes place on the Web, through a call centre, or in person, by determining the right mode, a wide assortment of much appreciated services can be placed at a person's fingertips.

Notes

1   Alan Westin, *Privacy and Freedom,* New York: Atheneum, 1967, p. 32.
2   1980 OECD *Guidelines Governing the Protection of Privacy and Trans-border Flows of Personal Data*
   http://oecd.org/dsti/sti/it/secur/prod/PRIV-EN.HTM

8

Larger Possibilities

I t's easy – and titillating – to focus on the dark side that could spread as corporations gain access to increasing amounts of personal information, but it is difficult to recognize in its full spectrum the potential of technology to accommodate the needs of people. Throughout this book, I've tried to balance the benefits of personalization with valid concerns about privacy, as well as lofty new strategies with basic practicality. But now it's time to be a little less even-handed and to explore the remarkable possibilities of personalization correctly done. The trends and forces at work are in their infancy, and their benefits could be much more far-reaching than simply stronger relationships between a company and its key stakeholders.

If any one of these four visions becomes reality, our world will be dramatically better:

- personalization begins not with companies, but with individuals themselves;
- companies that add value via personalization tread lighter on this planet and in the communities they impact;

- personalization increases diversity, which makes companies, and communities, less vulnerable to economic disasters;
- corporations become more accountable for their actions, especially with regard to indirect stakeholders.

Attend a personalization conference today, and you'll probably hear about marketing, and not much else. But marketing is 10% or less of the story. Personalization is about people, and the things that matter most to them, which, as we've already discussed, progresses from basic to higher order needs. The ultimate pay-off of personalization has to be that it enables us to help each other, because that's the behaviour that emerges at the top of the ladder.

When you consider the changes that personalization will bring to companies and to our economies, you first need to recognize that a company has at least two kinds of stakeholders: *direct* ones with whom it conducts direct financial transactions, and *indirect* ones with whom it does not. The first group includes customers, employees, partners, and suppliers. The second includes citizens of the communities in which a business is located, people who may live 100 miles downstream from a manufacturing plant but whose water quality is impacted by the plant's operation, and school officials in a neighbouring county from a large business who have to cope with rising enrolments thanks to the firm's expansion but who don't collect tax revenues from the business itself. The main difference between these two groups is that a firm can more easily quantify the value of its relationships with the first group, and so can those stakeholders. With the second group, the relationships are indirect and often involuntary. Profit motive alone can be enough to justify stronger relationships with the first group; it takes social responsibility to recognize the needs of the second.

Ultimately, personalization will make it easier for firms to exhibit what indirect stakeholders view as socially responsible behaviour. The practices and capabilities that underlie it will also make it easier for these stake-holders to quantify a company's impact on their interests.

Personalization Begins with Individuals

P2P community newswire

Highest popularity message (98% approval rating)

Keywords: trust, sincerity, reliability, proven, dependable

FROM: Paul Marion

DATE: 30 May 2004

RE: message for new participants

In 2002, I started this P2P (peer-to-peer) node to rate companies specifically with regard to a single question: do they do exactly what they say? In my mind, this is a fairer standard that any absolute measure of product quality or delivery time. Sometimes, you don't need – or want to pay for – the highest level of quality. Also, there can be good reasons why it takes a long time to deliver a product; I'd gladly wait 14 weeks to get a handmade bookcase rather than one produced on an assembly line.

We now have 14 million consumers worldwide contributing ratings, and we've managed to keep this node entirely voluntary and non-commercial. No one, including me, spends more than 10 hours a week here. I know, because we enforce this limit to prevent abuse.

Over the months, our system evolved into its current, simple format. Participants have only two choices when rating a company: yes (they do what they say) or no (the company does not do what they say).We consider a negative or positive rating statistically significant when 90% or more of our participants agree and when a sufficient number of ratings have been received (this level changes by company size; click here for details). At present, 532,190 companies have been rated. Of these, 43,117 have negative ratings and 14,320 have positive ratings. The balance are either inconclusive or too early to report.

The trend, however, is positive. Since we hit the 5 million participant level last November, companies have begun to react to negative ratings, as well as to publicize positive ones. Participants have reported increased efforts on the part of many companies to honour their word, and 412 products have been pulled off the market or had their claims restated as a direct result of our collective activities.

Your rating makes a big difference, so please consider your submissions carefully. Our goal is not to slam companies, nor is it to falsely inflate the reputations of those in whom we have a vested interest. If in doubt, don't vote. Of course, rules of this node prohibit anyone from rating a company in which he or she has a vested interest (employee, partner, supplier, investor); those who violate this rule will be banned from all P2P nodes (click here for the ten principles of P2P fairness adopted 2 January 2003).

Until now, individuals haven't had any real economic power without a company to bind them together. But the advent of peer-to-peer networks and the advancement of technology in general make it likely that individuals will group together to personalize the way they treat companies. These groups will be fluid and probably fiercely independent, since they will lack any sort of corporate interest at their centre and thus lack the sort of vested interests and basic stability that typifies companies.

On many levels, Napster's battles with the music industry and the courts remind us how important it is for business executives to broaden their thinking when considering the potential importance of personalization. The popular music-sharing service brought "peer to peer" – which means that each user's PC becomes a server in a decentralized network – into the popular vernacular. Napster's software helps users find files they want on the machines of other users, but Napster doesn't store any files itself. It rattles the common notion of a company as a hub of a wheel. To overcome the music industry's successful challenges to Napster, other services have emerged that are even more decentralized, eliminating any sort of central server at all. From a technology perspective, little prevents people from communicating directly with each other, in vast numbers, without the assistance of any corporation. All that's needed are freely distributed software and rallying points around which communities will gather.

I'm not talking about the sort of communities that exist today online, which already demonstrate the breadth of topics about which people feel strong enough to band together with others they do not yet know: guns, environmental protection, bargain hunting, porcelain dolls, education, skiing, reading, Dilbert, Harley Davidson, pure-bred German shepherds, farming, sailing, astronomy, heart disease, stamp collecting, or the music group Phish. Most of today's Web communities lack the capabilities that will empower individuals to automate their likes and dislikes, to swap files and opinions directly with other users that they have never met, or to search for and analyse data in large quantities. But eventually, perhaps soon, people will have these capabilities and will be able to interact with millions of others without the assistance of a corporation. This will mean that each person can pursue his own agenda, free from the influence of a company with its own agenda.

No matter how noble an organization's stated goals, groups change once they acquire significant operating budgets and full-time staffs. "Customer-driven" companies become less so when they hit a rough financial quarter and investors push them to cut costs. Even the leadership of charities can become entrenched and lose sight of the populations they are there to serve.

But true peer-to-peer networks of individuals, completely independent of any static organization, could escape many of these problems. Such networks are driven by standards and protocols, not boards of directors and executives. There are plenty of programmers with the talent and in-clination to develop software that lets individuals find each other and communicate, without any corporation footing the bill or influencing the outcome. This doesn't mean that plenty of corporations won't attempt to profit by building and nurturing such networks; it just means they won't be the only ones, and such networks may want to be free of corporate influence. Truly independent networks may be more unruly and chaotic than corporate alternatives, but that means they'll adapt faster to changing conditions. Nodes will spring to life – or die – in minutes or seconds, instead of the months and years it takes to create companies.

Depending on the needs or interests of each group, individuals could instruct their computers to always – or never – do business with companies that meet their standards. These standards could be based on the collective opinions or experiences of other members of the group. In Paul Marion's trust-based network, I would expect that participants would never do business with a company rated as untrustworthy. But soon technology will allow individuals to create business rules with just as much sophistica-tion as companies. A practical person might tell her shopping software to never do business with an "untrustworthy" company unless both of the following are true:

1. no merchant with better trustworthy ratings have the product; and
2. "85% or more" of consumers reported that they eventually received satisfaction from the untrustworthy merchant.

The ability to set these kinds of conditions – to make computers work for you, instead of the other way around – is what's still missing from the Web and other interfaces that most people can access today. But it's coming. When that happens, the balance of power will shift and individuals will have the option of putting themselves at the centre of personalization.

Tread Lightly

Personalization makes improving service more profitable than adding products. Driven by data and intelligence, rather than just physical goods, personalization lessens our drive to consume natural resources. We are early in a shift from manufacturing to services and eventually to robust personalization.

Figure 8.1   Main source of added value

In its 2000 annual report, General Electric highlighted its transformation from a manufacturing firm to a service provider:

"A **Services** focus has changed GE from a company that in 1980 derived 85% of its revenues from the sale of products to one that today is based 70% on the sale of services. This extends our market potential and our ability to bring value to our customers."

Many other companies are making a similar transition, and with good reason. In the manufacturing world, life is a constant struggle to avoid competing on price or being viewed as a commodity producer. Most manufacturing processes are still fairly inflexible and require large investments. Many manufacturers are disconnected from their customers, since they sell through distributors. This makes them vulnerable to "switching" tactics by middlemen. Significantly, many manufacturing processes are also fraught with environmental issues that can boomerang on a company months or years later. Over 30 years, for example, GE plants in upstate New York used polychlorinated biphenyls, or PCBs, as insulating fluid in capacitors. Twenty years ago, the company stopped using PCBs at its plants in Hudson Falls and Fort Edward, New York, but the company is still involved in efforts to clean up PCBs that remain in the Hudson River, which flows from upstate down to New York City. The company is in the news again, because the Environmental Protection Agency is proposing to make the company dredge the river at a cost of $460 million. GE is promoting an alternative $20 million to $30 million onshore source-control project.[1] The EPA supports onshore source control, but argues that such isn't enough to produce the needed clean-up of 100,000 pounds of PCBs buried in sediment. When GE was setting prices for its capacitors in 1960, you can bet they didn't include the cost of cleaning up the Hudson River over 20 years or running advertising campaigns designed to overcome the EPA's vastly more expensive clean-up efforts. Who knows when these costs will stop boomeranging back at the company? Such are the liabilities of manufacturing operations.

Services, in contrast, are much easier to customize and adapt, to suit both changing market and individual needs. Still, "services" is a generic term, and services can be just as undifferentiated as any mass-production process. Merely moving into a service business doesn't enable a company to lock in the loyalty of its stakeholders or even to lessen its need to consume natural resources. Since no added value comes from a company's past knowledge of a person, the company can only generate revenue by doing something. But a company that leverages personalization can get paid for doing nothing.

Michael Saylor, CEO of the software firm Microstrategy, told a story about an imaginary website called MyHouseIsOnFire.com that illustrates this point. The way the site works, said Saylor, is that it is connected to all the emergency-service databases around the world and you can go there periodically and type in your address. It then searches to see if your house is on fire, and if it isn't, then the site says, "Congratulations! Your house is not on fire. Please check back tomorrow." The site doesn't remember you, and if there is anyone in the world willing to pay for such a service, they certainly would only pay for the (few) times they actually used the site. But, said Saylor, imagine the same capabilities used differently. You sign up for a service – let's call it Peace of Mind – that constantly searches emergency databases, and if it ever finds a mention of your home address or any member of your family, it immediately contacts you in the fastest manner possible. This service requires no initiative on your part, and you might go years without interacting with it, other than to update your contact information. But it's worth far more to most people than MyHouseIsOnFire.com

From an economic perspective, Peace of Mind has a negligible marginal cost from adding another customer. The company is already searching every database; all it has to do is add a few queries to include your information. The process of serving a customer generates virtually no physical waste and causes zero pollution. It also has the ideal characteristics of a Learning Relationship: the more information a customer enters in terms of contact information, physical assets, and family members, the more convenient it becomes for him or her to remain loyal to the service. Switching to another comparable service would be more work than it's worth, so customers aren't as price sensitive as they might otherwise be.

The further to the right a company advances on the added-value spectrum, the greater its profit potential and the less its exposure to environmental disasters that can follow behind manufacturing operations. By saving stakeholders time or money, by helping them find better information, or by customizing a service to meet their particular needs, the company adds value in ways that often are intangible. This adds up to an enterprise that treads more lightly on this planet and in the communities in which it operates, making it easier for the firm to balance its need for growth and profits with its desire to be socially responsible. As long as such firms are cautious to respect individual privacy, their potential liabilities remain significantly lower than when they are actually manufacturing products.

Think about this from a personal perspective. Instead of paying for an object, you are paying for a continuous stream of benefits that come from a relationship that serves your interests. If you are an employee, you benefit from a relationship that offers you more than just a job. Even as an investor, you could have the opportunity to not just make a sound investment, but to derive significant other benefits as well.

Still, somebody has to manufacture products, right? Even in manufacturing situations, personalization minimizes the impact of such operations. By customizing products in response to a customer's specifications, a company can lower inventory levels and reduce waste, both of which lessen the demand on natural resources and potentially reduce pollution. Today, many companies make the best possible guess at how many products they can sell, then go ahead and produce that number. Then the products sit on shelves, waiting to be purchased. The ideal situation would be to make the product in response to an order. Simply by adopting mass customization – the process of modularization that makes large-scale personalization possible – companies can reduce waste and shorten lead times. This is because mass customization makes manufacturing processes more nimble and provides greater opportunities to use or reuse each component and capability.

Since these benefits are so far-reaching, society is likely to profit in a much more significant manner from personalization than from other attempts by industry to be socially responsible. For example, philanthropy is a popular mechanism that companies use to support the communities in which they operate. But, in recent years it has frequently become a means of supporting points of view and causes in which the company has a shared interest or ones that serve to counterbalance negative perceptions of a firm. In some cases, it would take an unattainable amount of philanthropy to balance a company's negative impact on society. Some tobacco companies, for example, have been generous in their philanthropic contributions, but many perceive they do this because they are so widely demonized for their distribution of products that cause people to become sick and die. Many people would gladly endorse the end of the tobacco industry's philanthropic efforts, if that meant the industry would simply stop producing tobacco.

Does this mean the tobacco industry can solve its problems by personalizing cigarettes? No, because the basic product still has significant flaws. Remember the Corporate Needs Ladder? I'd argue that causing significant portions of a customer base to become ill as a result of using the product represents a notable quality flaw. But such companies might consider following General Electric's example and migrating to different types of businesses that are better suited to reap the benefits I'm discussing.

Diversity vs. Efficiency

The mass-production model still drives most businesses today, because it provides cost efficiencies that enable businesses to grow and be profitable. But it also can result in a degree of uniformity that makes a firm susceptible to economic disasters. This vulnerability applies to all types of business, but is easiest to spot in businesses that are related to nature.

The public radio programme *Marketplace* broadcast a commentary by William Halweil in which he argued that the most recent outbreak of foot-and-mouth disease in Europe was both inevitable and encouraged, from an animal-husbandry standpoint, by the poor livestock living conditions of industrial farming.[2] In the industry's quest for efficiency, it houses large masses of animals in closed quarters in which illness and other problems can spread rapidly. It then processes meat in similarly centralized locations. Bad feed or the outbreak of such a disease can then spread far more rapidly than if animals and meat were in less proximity to each other. Viewed from another perspective, an unnatural concentration of any single species provides an ideal situation for any disease that thrives on that species. Diseases that might otherwise never cause significant harm can be devastating when they are introduced to a population that lacks diversity, whether the operation concentrates animals, trees, rice, or any other crop.

While the economic benefits of diversity are well recognized, most industries have been forced to ignore many of these benefits in order to satisfy the demand for their products and their need to be profitable. Doing so exposes them to significant risks, such as any unanticipated and significant shift in the needs of their customers, the state of their supply chain, or the environment – economic, social, or natural – in which they operate. Personalization offers the first practical escape from this conundrum, by making diversity both profitable and scaleable.

Diversity doesn't just apply to the products a firm produces, but instead to the full scope of a company's activities. By having a culturally diverse workforce, companies are more likely to spot and exploit opportunities since they benefit from a wider range of perspectives and knowledge. Diverse methods of production provide insurance against technical and process problems. Diverse strategies and investment plans keep a company's options open and increase the odds that it will profit from inevitable but unexpected changes.

At its heart, personalization requires a 180-degree change in business thinking from "let's profit by convincing people their needs are similar" to "let's profit by encouraging people to be unique". Only people who believe their needs are unique have reason to be loyal to the company that has best learned to serve those needs. This new thought process values diversity, profits from it, and ultimately creates a profoundly healthier, safer, and more sustainable way of doing business.

Once executives – and the businesses they lead – start to make this shift, changes that were nearly impossible to accomplish before will become high-level priorities. Just as with social responsibility, diversity in the workforce has been a "nice to have" to the extent that a company has been able to exceed the legal requirements. Companies would like to offer opportunities to all, but doing so isn't their main business and it isn't how they make a profit. Remember, most companies profit from individual similarities, not differences, and thus many executives don't see how they are able to profit tangibly from diversity of opinions, attitudes, and backgrounds. But when a company decides that to thrive it must profit from individual differences, suddenly it needs to have diverse employees. Diversity adds a lot of value.

Likewise, to maximize the value of its relationship with an individual, a company will need a diverse set of capabilities and services. As companies shift from a manufacturing/product focus to one driven by service and personalization, it becomes more important for them to lock in the loyalty of the most valuable stakeholders, rather than to do business with as many stakeholders as possible regardless of their value. This means doing more for less people. Again, diversity becomes a must-have.

These problems and opportunities are interrelated. Business models based on growth through mere volume are ultimately unsustainable for companies, communities, and our planet. They require executives to minimize the needs of indirect stakeholders to achieve goals demanded by investors, customers, and employees. Just as mass marketing is an idea whose time has passed, so, too, must mass production give way to a more sustainable, more mutually beneficial business approach. We are all different, but we have common interests and are social creatures. Over the coming months and years, it will become increasingly possible for businesses and individuals to establish and support win–win relationships that benefit not only the direct parties to that relationship, but also respect the needs of other stakeholders in close proximity.

Corporate Accountability

It's not just information about individuals that is multiplying like rabbits in corporate and government databases. It's also information about companies, and this data will make it much easier for individuals and governments – as well as other companies – to track the behaviour of enterprises and hold them accountable for their actions. In fact, the mere absence of data in certain situations will be enough to sound alarms that will alert regulators and/or legislators. This trend will force managers to make a decision: do they view this increased accountability as a new weapon in the hands of indirect stakeholders or as a valuable tool that enables them to gain better control of the full scope of their companies? The question comes down to trust, and the latter path is the one that builds it.

In the environmental space, Greenpeace developed a four-point "CARE" checklist designed to separate green companies from those engaging in "greenwash". CARE stands for Core business, Advertising record, Research and development funding, and Environmental lobbying. Core business means that if a firm's primary business is based on an activity that causes harm to the environment, people need to be sceptical of its claims to be environmentally responsible. Companies use advertising to alter the public's perception of their brand and their activities, which is a perfectly acceptable practice as long as such promotions accurately reflect the true state of their activities. "All over the world Shell is advertising their green credentials yet their renewable energy investment is minuscule at 0.6 per cent of their total annual investment; Shell will have to do much more to convince shareholders and consumers that it is serious about tackling climate change," said a Greenpeace International energy solutions campaigner, Karl Mallon.[3] R&D funds provide a means to judge whether a firm is using such expenditures in a manner designed to solve problems that cause significant concern to stakeholders. It's not useful to learn that a firm has invested £500 million in R&D unless you know how much of this amount has been devoted to reducing pollution or some other mutually beneficial outcome. Environmental lobbying is a measure of whether a company puts its money someplace other than where its mouth is, by lobbying against legislation that its other activities might suggest it would support.

In the past, the problem with watchdog groups was that they tended to be self-defeating, by not being organized or disciplined enough and sometimes by offending people who might otherwise be sympathetic to their cause. But this is changing. Greenpeace, which gained notoriety through stunts such as members' chaining themselves to oil platforms, in 2000 took, for it, the radical step of buying enough shares in Shell that the firm's bylaws allowed Greenpeace to present a proposal that Shell build a large-scale solar-panel factory. To support its case, Greenpeace even hired the consulting firm KPMG to explore the feasibility of such a plant.

I'm an optimist at heart and would like to believe that we will see companies and their most vocal stakeholders, however indirect, being pushed to a middle ground by the forces that enable personalization. Emotion has its place, but to justify financial decisions most firms need the ability to quantify the results of their investments. Growing volumes of data make it easier for both companies and their indirect stakeholders to provide exactly this type of quantification. The data will allow people outside companies to report with increasing detail exactly how a company has spent its advertising and R&D, and to compare such investments to similar firms. While companies will always try to keep details about their investments confidential, no organization is immune from the impacts of advancing technologies. It won't be long before Greenpeace and hundreds of other large advocacy groups are hiring database analysts to shift through terabytes of data to find previously hidden patterns. Although pieces of the puzzle remain hidden, there will be plenty that reveal the bulk of any company's activities.

The companies that will be impacted first and most significantly by this trend will be those that already command the attention of activists, but eventually every company large enough to anger a resourceful person will have to take notice. Today, such a person can only create a [Your

Company's Name] Sucks.com website. Soon, that person will be able to search databases and publicize potentially embarrassing data, or even more significantly, demonstrate to others how they can use the knowledge to alter their own actions.

Ironically, personalization makes each of us increasingly responsible not just for our own personal welfare, but for the welfare of others. When we get more choice, we face a decision about how to best exercise it. If we use it solely to benefit ourselves, we are no better than a company that ignores its indirect stakeholders. Fortunately, most people when given a choice will help others if they have the ability to do so.

Many believe that the "generation Y" who are entering the workforce now are more disenfranchised and less eager to buy into the way that business has been done over the past decades. In my observations, employees of this generation, who are most likely to be knowledge workers, are the ones who are the least motivated simply by money or career success. As Peter Drucker, the author of more than 15 management books, wrote in *Atlantic Monthly*, "when (motivation) can no longer be done by satisfying knowledge workers' greed, it will have to be done by satisfying their values, and by giving them social recognition and social power." If a firm wants to deliver personalization to people who do not value money alone but who value highly social responsibility or sustainable business models, then somehow it has to honour these wishes to maintain the relationships.

Congratulations, You're Famous

It's not just companies that will be increasingly accountable for their actions. You will be, too. To get this message across, I've been giving a speech that connects celebrities, your privacy, and your life. The talk goes something like this: soon we will all be living like celebrities do today, seemingly in glass houses with prying eyes all around. Just as is the case with celebrities, our salaries, vacations, children, travels, injuries, missteps, reviews … will all be there for others to see. Very little, if anything, may remain private.

Although I use light-hearted examples, this is a troubling message for most who hear it. But, using examples from celebrity lives, I point out that even people who live under such constant scrutiny find ways to live enjoyable lives. They find agents they can trust, seek out merchants who offer them both excellent service as well as discretion, and accept responsibility for their actions. (Translation: only if this movie opens big do they get more money for the next one.)

Despite all the ground we've covered, this analogy makes it possible to summarize in two paragraphs what I hope you take away from this book.

In your work life, think of your business as a trusted agent and discrete partner, capable of using information about others to *their* benefit and of earning a fair profit for your discretion and expertise. Leverage technology to build stronger, more profitable business relationships with others.

In your personal life, look for those firms that treat you as a celebrity expects to be treated. You are unique, and your needs are unique. Do not settle for cookie-cutter service. Your right to privacy and personalized service may not yet be legislated, but any company that expects to profit in coming years will need to offer people like you precisely these types of services.

Thought Exercise . . .

What Can You Do Today?

Business will continue to become more personal. Computers, with their ability to collect, store, and remember information, will become increasingly pervasive in our lives. Armed with your own experience and the ideas collected here, what could you do today to better balance personalization, privacy, and profits? Success in this area provides a rare opportunity for motivated people like you to enjoy personal rewards while simultaneously helping many others around you. All it takes is initiative.

Notes

1 "GE Set For Hudson Clean-up," WCBS Newsradio 880, 9 March 2001
   http://cbsnewyork.com/nycMarket/news88/StoryFolder/story 062201135 html
   and HudsonVoice, a GE site at
   http://207.141.150.134/

2 *Marketplace*, 7 March 2001.

3 "Greenpeace Calls on Shell at AGM to Build," from Greenpeace press releases, 10 May 2000
   <James.Williams@ams.greenpeace.org>

Index